Lei Zhi Chen, Sing Kiong Nguang, Xiao Dong Chen

Modelling and Optimization of Biotechnological Processes

Studies in Computational Intelligence, Volume 15

Editor-in-chief
Prof. Janusz Kacprzyk
Systems Research Institute
Polish Academy of Sciences
ul. Newelska 6
01-447 Warsaw
Poland
E-mail: kacprzyk@ibspan.waw.pl

Lei Zhi Chen
Sing Kiong Nguang
Xiao Dong Chen

Modelling and Optimization of Biotechnological Processes

Artificial Intelligence Approaches

 Springer

Dr. Lei Zhi Chen
Diagnostics and Control Research Centre
Engineering Research Institute
Auckland University of Technology
Private Bag 92006, Auckland
New Zealand
E-mail: lezhi.chen@aut.ac.nz

Professor Dr. Xiao Dong Chen
Department of Chemical
and Materials Engineering
The University of Auckland
Private Bag 92019, Auckland
New Zealand
E-mail: d.chen@auckland.ac.nz

Professor Dr. Sing Kiong Nguang
Department of Electrical
and Computer Engineering
The University of Auckland
Private Bag 92019, Auckland
New Zealand
E-mail: sk.nguang@auckland.ac.nz

ISSN print edition: 1860-949X
ISSN electronic edition: 1860-9503
ISBN:13 978-3-642-06792-1
e-ISBN:13 978-3-540-32493-5

Springer is a part of Springer Science+Business Media
springer.com
© Springer-Verlag Berlin Heidelberg 2006
Softcover reprint of the hardcover 1st edition 2006

is especially promising when it is costly or even infeasible to gain *a prior* knowledge or detailed kinetic models of the processes.

Auckland   *Lei Zhi Chen*
October, 2005   *Sing Kiong Nguang*
                                                         Xiao Dong Chen

Preface

Most industrial biotechnological processes are operated empirically. One of the major difficulties of applying advanced control theories is the highly nonlinear nature of the processes. This book examines approaches based on artificial intelligence methods, in particular, genetic algorithms and neural networks, for monitoring, modelling and optimization of fed-batch fermentation processes. The main aim of a process control is to maximize the final product with minimum development and production costs.

This book is interdisciplinary in nature, combining topics from biotechnology, artificial intelligence, system identification, process monitoring, process modelling and optimal control. Both simulation and experimental validation are performed in this study to demonstrate the suitability and feasibility of proposed methodologies. An online biomass sensor is constructed using a recurrent neural network for predicting the biomass concentration online with only three measurements (dissolved oxygen, volume and feed rate). Results show that the proposed sensor is comparable or even superior to other sensors proposed in the literature that use more than three measurements. Biotechnological processes are modelled by cascading two recurrent neural networks. It is found that neural models are able to describe the processes with high accuracy. Optimization of the final product is achieved using modified genetic algorithms to determine optimal feed rate profiles. Experimental results of the corresponding production yields demonstrate that genetic algorithms are powerful tools for optimization of highly nonlinear systems. Moreover, a combination of recurrent neural networks and genetic algorithms provides a useful and cost-effective methodology for optimizing biotechnological processes.

The approach proposed in this book can be readily adopted for different processes and control schemes. It can partly eliminate the difficulties of having to specify completely the structures and parameters of the complex models.It

Contents

1

Introduction

1.1 Fermentation Processes

Fermentation is the term used by microbiologists to describe any process for the production of a **product** by means of the mass culture of a **microorganism** [1]. The **product** can either be: i) The cell itself: referred to as biomass production. ii) A microorganism's own metabolite: referred to as a product from a natural or genetically improved strain. iii) A microorganism foreign product: referred to as a product from recombinant DNA technology or genetically engineered strain.

There are three types of fermentation processes existing: batch, continuous and fed-batch processes. In the first case, all ingredients used in the bioreaction are fed to the processing vessel at the beginning of the operation and no addition and withdrawal of materials take place during the entire batch fermentation. In the second case, an open system is set up. Nutrient solution is added to the bioreactor continuously and an equivalent amount of converted nutrient solution with microorganisms is simultaneously taken out of the system. In the fed-batch fermentation, substrate is added according to a predetermined feeding profile as the fermentation progresses. In this book, we focus on the fed-batch operation mode, since it offers a great opportunity for process control when manipulating the feed rate profile affects the productivity and the yield of the desired product [2]. A picture of laboratory bench-scale fermentors is shown in Figure 1.1. The schematic diagram of the fed-batch fermentor and its control setup is illustrated in Figure 1.2.

Fermentation processes have been around for many millennia, probably since the beginning of human civilization. Cooking, bread making, and wine making are some of the fermentation processes that humans rely upon for survival and pleasure. Though they link strongly to human daily life, fermentation processes did not receive much attention in biotechnology and bioengineering research activities until the second half of the twentieth century [3].

An important and successful application of fermentation process in history is the production of penicillin [4]. In 1941, only a low penicillin productivity of

L. Z. Chen et al.: *Modelling and Optimization of Biotechnological Processes*, Studies in Computational Intelligence (SCI) **15**, 1–16 (2006)
www.springerlink.com

Fig. 1.1. Laboratory bench-scale fermentation equipment used in the research. Model No.: BioFlo 3000 bench-top fermentor. Made by New Brunswick Scientific Co., INC., USA.

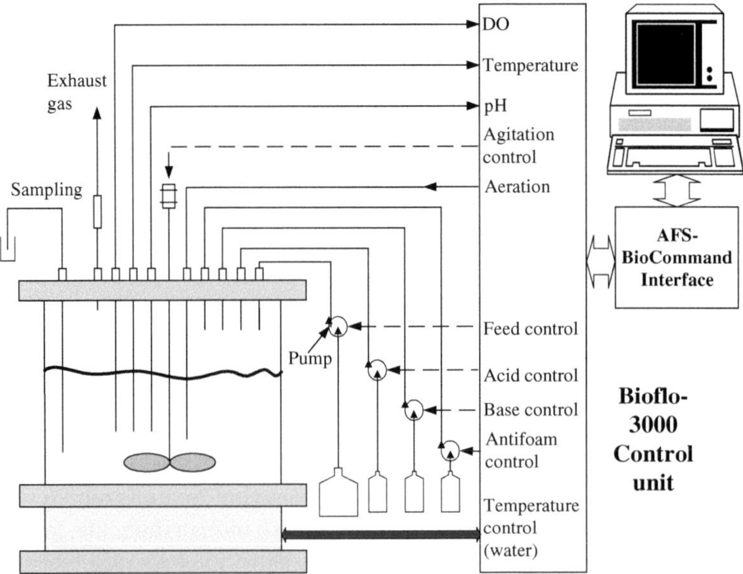

Fig. 1.2. Schematic diagram of the computer-controlled fed-batch fermentation.

about 0.001 g/L could be obtained by surface culturing techniques, even when high-yielding strains were used. The demand for penicillin at that time exceeded the amount that could be produced. In 1970, the productivity was dramatically increased to over 50 g/L by well-controlled large-scale, submerged and aerated fermentation. As a result, more human's lives were saved by using penicillin. Since then, a large number of innovative products, such as specialty chemicals, materials for microelectronics, and particularly, biopharmaceuticals, have been manufactured using fermentation processes and have been making a significant contribution in improving health and the quality of life [1]. The twenty first century is thus regarded as "the biotechnology century".

Although fermentation operations are abundant and important in industries and academia which touch many human lives, high costs associated with many fermentation processes have become the bottleneck for further development and application of the products. Developing an economically and environmentally sound optimal cultivation method becomes the primary objective of fermentation process research nowadays [5]. The goal is to control the process at its optimal state and to reach its maximum productivity with minimum development and production cost, in the mean time, the product quality should be maintained. A fermentation process may not be operated optimally for various reasons. For instance, an inappropriate nutrient feeding policy will result in a low production yield, even though the level of feed rate is very high. An optimally controlled fermentation process offers the realization of high standards of product purity, operational safety, environmental regulations, and reduction in costs [6].

Though many attempts have been made in improving the control strategies, the optimization of fermentation processes is still a challenging task [7], mainly because:

- The inherent nonlinear and time-varying (dynamic) nature make the process extremely complex.
- Accurate process models are rarely available due to the complexity of the underlying biochemical processes.
- Responses of the process, in particular for cell and metabolic concentrations, are very slow, and model parameters vary in an unpredictable manner.
- Reliable on-line sensors which can accurately detect the important state variables are rarely available.

1.2 Fed-Batch Fermentation Processes by Conventional Methods

Process monitoring

Process monitoring, which is also called state estimation, is very important for implementation of on-line control strategies [8]. Dissolved oxygen (DO) and pH are the most commonly measured parameters using electrochemical sensors [9]. However, some key state variables, such as biomass concentration, may not be measured directly due to the lack of suitable sensors or high costs. In recent years, lots of efforts have been involved in on-line software sensor (softsensor) development. The key concept of softsensor techniques is to estimate unmeasured states from measured states. Unmeasured states are normally inaccessible or difficult to measure by means of biosensors or hardware sensors, while measured states are relatively easy to monitor on-line using reliable well-established instruments. Based on this philosophy, several softsensor techniques have been proposed in the literature [10], namely:

- estimation using elemental balances [11];
- adaptive observer [12];
- filtering techniques (Kalman filter, extended Kalman filter) [13].

The first two methods suffer from the inaccuracies of available instruments and models. The third method requires much design work and prior estimates of measurement noise and model uncertainty characteristics. It also suffers from some numerical problems and convergence difficulties due to the approximation associated with model linearization.

Process modelling

The key of the optimal control problem is generally regarded as being a reliable and accurate model of the process. For many years, the dynamics of bioprocesses in general have been modelled by a set of first or higher order nonlinear differential equations [14]. These mathematical models can be divided into two different categories: structured models and unstructured models. Structured models represent the processes at the cellular level, whereas unstructured models represent the processes at the population (extracellular) level.

Lei et al. [15] proposed a biochemically structured yeast model, which was a moderately complicated structured model based on Monod-type kinetics. A set of steady-state chemostat experimental data could be described well by the model. However, when applied to a fed-batch cultivation, a relatively large error was observed between model simulation and the experimental data. Another structured model to simulate the growth of baker's yeast in industrial bioreactors was presented by Di Serio et al. [16]. The detailed modelling of regulating processes was replaced by a cybernetic modelling framework, which was based on the hypothesis that microorganisms optimize the utilization of available substrates to maximize their growth rate at all times. From the simulation results that were plotted in the paper, the model prediction agreed reasonably well with both laboratory and industrial fed-batch fermentation data that were adopted in the study. Unfortunately, detailed error analysis neglected to show what degree of accuracy could be achieved by the model. The limitation of the model, as pointed out by the authors, was that the model and it's parameters needed to be further improved for a more general application.

A popular unstructured model for industrial yeast fermenters was reported by Pertev et al. [17]. The kinetics of yeast metabolism, which were considered to build the model, were based on the limited respiratory capacity hypothesis developed by Sonnleitner and Käppeli [18]. The model was tested for two different types of industrial fermentation (batch and fed-batch modes). The results showed that it could predict the behaviors of those industrial scale fermenters with a sufficient accuracy. Later, a study carried on by Berber et al. [19] further showed that by making use of this model, a better profile of substrate feed rate could be obtained to increase the biomass production, while in the mean time, decreasing the ethanol formation. Recent application of the model has been to evaluate various schemes for controlling the glucose feed rate of fed-batch baker's yeast fermentation [20]. Because intracellular state variables (i.e., enzymes) are not involved in unstructured models, it is relatively easy to validate these kinds of models by experiments. This is why unstructured models are more preferable than structured models for optimization and control of fermentation processes. However, unstructured models also suffer the problems of parameter identification and large prediction errors.

The parameters of the model vary from one culture to another. Conventional methods for system parameter identification such as Least Squares, Recursive Least Squares, Maximum Likelihood or Instrument Variable work well for linear systems. Those schemes, however, are in essence local search techniques and often fail in the search for the global optimum if the search space is not differentiable or is nonlinear in parameters.

Though a considerable effort has been made in developing detailed mathematical models, fermentation processes are just too complex to be completely described in this manner. "The proposed models are by no means meant to mirror the complete yeast physiology ..." [15]. From an application point of view, the limitations of mathematical models are:

- Physical and physiological insight and *a priori* knowledge about fermentation processes are required.
- Only a few metabolites can be included in the models.
- The ability to cope with batch to batch variations is poor .
- These models only work under idea fermentation conditions.

- A high number of differential equations (high order system) and parameters are presented in the models, even for a moderately complicated model.

Process optimization

Systematic development of optimal control strategies for fed-batch fermentation processes is of particular interest to both biotechnology-related industries and academic researches [2, 7, 14], since it can improve the benefit/cost ratio both economically and environmentally. Many biotechnology-based products such as pharmaceutical products, agricultural products, specialty chemicals and biochemicals are made in fed-batch fermentations commercially. Fed-batch is generally superior to batch processing on the final yield. However, maintaining the correct balance between the feed rate and the respiratory capacity is a critical task. Overfeeding is detrimental to cell growth, while underfeeding of nutrients will cause starvation and thus reduce the production formation too. From the process engineering point of view, it opens a challenging area to maximize the productivity by finding the optimal control profile.

In reality, to control a fed-batch fermentation at its optimal state is not straightforward as mentioned above. Several optimization techniques have been proposed in the literature [7]. The conventional optimization methods that are based on mathematical optimization techniques are usually unable to work well for such systems [21]. Pontryagin's maximum principle has been widely used to optimize penicillin production [22] and biphasic growth of *Saccharomyces Carlsbergensis* [23]. The mathematical models used in all these cases are of low-order systems, i.e., a fourth order system. However, it becomes difficult to apply Pontryagin's maximum principle if a system is of an order greater than five.

Dynamic programming (DP) algorithms have been used to determine the optimal profiles for hybridoma cultures [24, 25]. For the fed-batch culture of hybridoma cells, more state variables are required to describe the culture since the cells grow on two main substrates, glucose and glutamine, and release toxic products, lactate and ammonia, in addition to the desired metabolites. This leads to a seventh order model for fed-batch operation, hence, it is difficult to apply Pontryagin's maximum principle. the DP is thus used to determine optimal trajectories for such high-order systems. However, the search space comprises all possible solutions to the high-order systems and is too large to be exhaustively searched. A huge computational effort is involved in this approach which sometimes may lead to a sub-optimal solution.

1.3 Artificial Intelligence for Optimal Fermentation Control

As early as the 1960s, artificial intelligence (AI) appeared in the control field, and a new era of control was born [26, 27]. Chronologically, expert systems, fuzzy logic, artificial neural networks (ANNs) and evolutionary algorithms (EAs), particularly genetic algorithms (GAs), have been applied to add "intelligence" to various control systems. Recent years have witnessed the rapidly growing application of AI to biotechnological processes [28, 29, 30, 31, 32, 33].

Each of the AI techniques offers new possibilities and makes intelligent control more versatile and applicable in an ever-increasing range of bioprocess controls. These approaches, in most part, are complementary rather than competitive. They are also utilized in combination, referred to as "hybrid". In this book, the combination of ANNs and GAs are used to optimize the fed-batch bioreactors.

A brief review of neural networks, GAs and their applications to biotechnological process controls is presented below. This helps to lay the groundwork for intelligent monitoring, modelling and optimal control of fed-batch fermentation described later in the book.

Recurrent neural networks: basic concepts and applications for process monitoring and modelling

ANNs are computational systems with an architecture and operation inspired from our knowledge of biological neural cells (neurons) in the brain. They can be described either as mathematical and computational models for static and dynamic (time-varying) non-linear function approximation, data classification, clustering and non-parametric regression or as simulations of the behavior of collections of model biological neurons. These are not simulations of real neurons in the sense that they do not model the biology, chemistry, or physics of a real neuron. They do, however, model several aspects of the information combining and pattern recognition behavior of real neurons in a simple yet meaningful way. Neural modelling has shown incredible capability for emulation, analysis, prediction, and association. ANNs are able to solve difficult problems in a way that resembles human intelligence [34]. What is unique about neural networks is their ability to learn by examples. ANNs can and should, however, be retrained on or off line whenever new information becomes available.

There exist many different ANN structures. Among them there are two main categories in use for control applications: feedforward neural network (FNN) and recurrent (feed back) neural network (RNN) [35, 36]. FNN consists of only feed-forward paths, its node characteristics involve static nonlinear functions. An example of a FNN is shown in Figure 1.3. In contrast to FNNs,

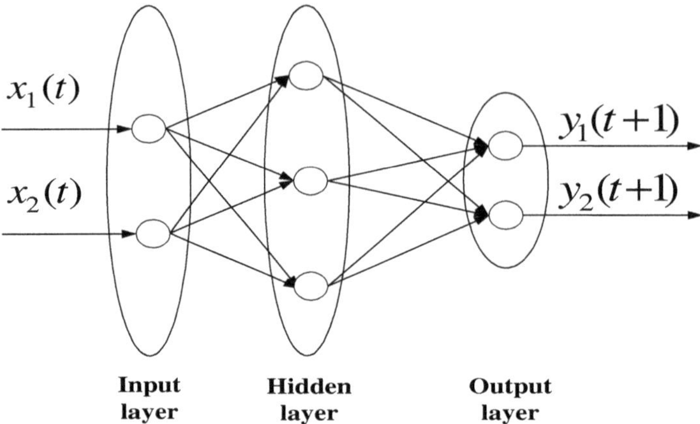

Fig. 1.3. Topological structure of a FNN.

the topology in RNNs consists of both feed-forward and feedback connections, its node characteristics involve nonlinear dynamic functions and can be used to capture nonlinear dynamic characteristics of non-stationary systems [7, 37]. An example of RNN is illustrated in Figure 1.4.

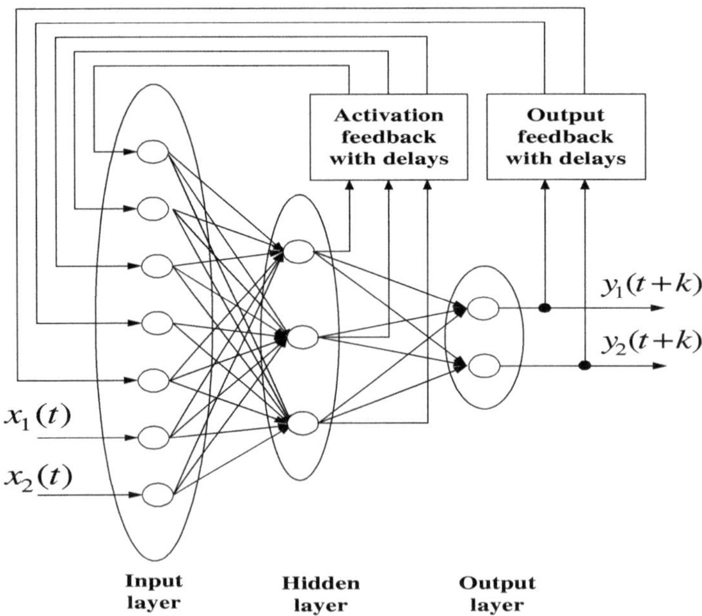

Fig. 1.4. Topological structure of a RNN.

Recurrent neural networks for state estimation

Some attempts have been made to estimate important states in batch and fed-batch bioreaction using RNNs. In the beginning, an RNN had either of two basic configurations - the Elman form or the Jordan form [37, 38]. The original purposes of these two networks were to control robots and to recognize speech. Later, due to their intrinsic dynamic nature, RNNs drew considerable attention in the research area of biochemical engineering [7]. An application of an Elman RNN to fed-batch fermentation with recombinant *Escherichia coli* was reported by Patnaik [39]. The Elman RNN was employed to predict four state variables in the case of flow failure. The performance of the RNN was found to be superior to that of the FNN network. Since both of the Elman and Jordan networks are structurally locally recurrent, they are rather limited in terms of including past information. A recurrent trainable neural network (RTNN) model was proposed to predict and control fed-batch fermentation of *Bacillus thuringiensis* [40]. This two layer network has recurrent connections in the hidden layer. the Backpropagation algorithm was used to train the network. The results showed that the RTNN was reliable in predicting fermentation kinetics provided that sufficient training data sets were available. In this research, RNNs with both activation feedback and output feedback connections are used for on-line biomass prediction of fed-batch baker's yeast fermentation.

A moving window, feed-forward, backpropagation neural network was proposed to estimate the consumed sugar concentration [41]. Since the FNN was primarily used for nonlinear static mapping, the dynamic nature of the fed-batch culture was imposed by the moving window technique. The data measured one hour ago was used to predict the current state. The oldest data were discarded and the newest data were fed in through the moving window method. In a new approach, the RNN was adopted to predict the biomass concentration in baker's yeast fed-batch fermentation processes [42]. In contrast to FNNs, the structure of RNNs consists of both feed-forward and feedback connections. As a result of feedback connections, explicit use of the past outputs of the system is not necessary for prediction. The only inputs to the network are the current state variables. Thus, the moving window technique is not necessary in this RNN approach for biomass concentration estimation.

Recurrent neural networks for process modelling

Neural networks as alternative tools have been extensively studied in process modelling because of their inherent capability to handle general nonlinear dynamic relationships between inputs and outputs. Many reviews of the applications of ANNs in modelling and control of biotechnological processes can be found in the literature [28, 30, 29, 31, 2, 43]. Neural networks are able to extract underlying information from real processes in an efficient manner with normal availability of data. The main advantage of this data-driven approach is that modelling of complex bioprocesses can be achieved without *a priori* knowledge or detailed kinetic models of the processes [44, 45, 46, 47, 48, 49].

RNN structures are more preferable than FNN structures for building bioprocess models, because the topology of RNNs characterize a nonlinear dynamic feature [50, 7, 51, 52, 53, 54, 55, 56]. The connections in RNNs include both feed-forward and feedback paths in which each input signal passes through the network more than once to generate an output. The storage of information covering the prediction horizon allow the network to learn complex temporal and spatial patterns. A RNN was employed to simulate a fed-batch fermentation of recombinant *Escherichia coli* subject to inflow disturbances [39]. The network that was trained with one kind of flow failure was used to predict the course of fermentation for other kinds of failures. It was found that the recurrent network was able to simulate the other two unseen processes with different inflow disturbances, and the prediction errors were smaller than those with FNNs for similar systems. Another comparison study was made by Acuña et al. [57]. Both static and recurrent (dynamic) network models were used for estimating biomass concentration during a batch culture. The dynamic model performed implicit corrective actions to perturbations, noisy measurements and errors in initial biomass concentrations. The results showed that the dynamic estimator was superior to the static estimator at the above aspects. Therefore, there is no doubt that the RNNs are more suitable than FNNs for the purpose of bioprocess modelling.

The prediction accuracy of the RNN models is heavily dependent on the structure being selected. The determination of the RNN structure includes the selection of the number of hidden neurons, the connection and the delays of feedback, and the input delays. It is problem specific and few general guidelines exist for the selection of the optimal nodal structure [28]. The above mentioned RNNs are structurally locally recurrent, globally feed-forward networks. These structures are rather limited in terms of including historical information [37], because the more feedback connections the RNNs have, the "dynamically richer" they are. A comparison between RNNs and augmented RNNs for modelling a fed-batch bioreactor was presented by Tian et al. [58]. The accuracy of long range prediction of secreted protein concentration was significantly improved by using the augmented RNN which contains two RNNs in series.

In this book, an extended RNN is adopted for modelling fed-batch fermentation of *Saccharomyces cerevisiae*. The difference between the extended RNN and the RNNs mentioned above is that, besides the output feedback, the activation feedbacks are also incorporated into the network, and tapped delay lines (TDLs) are used to handle the input and feedback delays. A dynamic model is built by cascading two such extended RNNs for predicting biomass concentration. The aim of building such a neural model is to predict biomass concentration based purely on the information of the feed rate. Therefore, the model can be used to maximize the final quantity of biomass at the end of reaction time by manipulating the feed rate profiles.

Genetic algorithms: Basic concepts and applications for model identification and process optimization

In this book, the idea of the biological principle of natural evolution (survival of the fittest) to artificial systems is applied. This idea was introduced more than three decades ago. It has seen impressive growth in application to biochemical processes in the past few years. As a generic example of the biological principle of natural evolution, GAs [59, 60, 61, 62, 63, 64, 65, 66, 67] are considered in this research. GAs are optimization methods, which operate on a number of candidate solutions called a "population". Each candidate solution of a problem is represented by a data structure known as an "individual". An individual has two parts: a chromosome and a fitness. The chromosome of an individual represents a possible solution of the optimization problem ("chromosome" and "individual" are sometimes exchangeable in the literature) and is made up of genes. The fitness indicates how well an individual of the population solves the problem.

Though there are several variants of GAs, the basic elements are common: a chromosomal representation of solutions, an evaluation function mimicking the role of the environment, rating solutions in terms of their current fitness, genetic operators that alter the composition of offspring during reproduction and values of the algorithmic parameters (population size, probabilities of applying genetic operators, etc). A template of a general formulation of a GA is given in Figure 1.5. The algorithm begins with random initialization of the population. The transition of one population to the next takes place via the application of the genetic operators: crossover, mutation and selection. Crossover exchanges the genetic material (genes) of two individuals, creating two offspring. Mutation arbitrarily changes the genetic material of an individual. The fittest individuals are chosen to go to the next population through the process of selection. In the example shown in Figure 1.5, The GA assumes user-specified conditions under which crossover and mutation are performed, a new population is created, and whereby the whole process is terminated.

GAs are stochastic global search methods that simultaneously evaluate many points in the parameter space. The selection pressure drives the population towards a better solution. On the other hand, mutation can prevent GAs from being stuck in local optima. Hence, it is more likely to converge towards a global solution. GAs mimic evolution, and they often behave like evolution in nature. They are results of the search for robustness; natural systems are robust - efficient and efficacious - as they adapt to a wide variety of environments. Generally speaking, GAs are applied to problems in which severe nonlinearities and discontinuities exist, or the spaces are too large to be exhaustively searched. As a summary, the general features that GAs have are listed below [69]:

- GAs operate with a population of possible solutions (individuals) instead of a single individual. Thus, the search is carried out in a parallel form.

Genetic algorithm

Choose an initial population of chromosomes;

while termination condition not satisfied **do**

   repeat

      if crossover condition satisfied **then**

      {select parent chromosomes;

      choose crossover parameters;

      perform crossover}

      if mutation condition satisfied **then**

      {select chromosome(s) for mutation;

      choose mutation points;

      perform mutation};

      evaluate fitness of offspring;

   until sufficient offspring created;

   select new population;

endwhile

Fig. 1.5. Structure of a GA, extracted from Fig. 2.2, Page 26 in [68].

- GAs are able to find optimal or suboptimal solutions in complex and large search spaces. Moreover, GAs are applicable to nonlinear optimization problems with constraints that can be defined in discrete or continuous search spaces.
- GAs examine many possible solutions at the same time. So there is a higher probability that the search converges to an optimal solution.

1.4 Why is Artificial Intelligence Attractive for Fermentation Control?

The last decade or so, has seen a rapid transition from conventional monitoring and control based on mathematical analysis to soft sensing and control based on AI. In an article on the historical perspective of systems and control, Zadeh considers this decade as the era of intelligent systems and urges for some tuning [70]:

> "I believe the system analysis and controls should embrace soft computing and assign a higher priority to the development of methods that can cope with imprecision, uncertainties and partial truth".

Fermentation processes, as mentioned in Section 1.1, are exceedingly complex in their physiology and performance. To propose mathematical models that are sufficiently accurate, robust and simple is a time-consuming and costly work, especially in the noisy interactive environment. AI, particularly neural networks, provides a powerful tool to handle such problems. An illustration of a neural network-based biomass and penicillin predictor has been given by Di Massimo et al. [71]. The neural network of relatively modest scale was demonstrated to be able to capture the complex bioprocess dynamics with a reasonable accuracy. The ability to infer some important state variables (eg. biomass) from other measurements makes neural networks very attractive in the applications of fermentation monitoring and modelling [72, 73, 74], because it can reduce the burden of having to completely construct the mathematical models and to specify all the parameters.

The dynamic optimization problems of such complex, time-variant and highly nonlinear systems are difficult to solve. The conventional analytical methods, such as Green's theorem and the maximum (or minimum) principle of Pontryagin, are unable to provide a complete solution due to singular control problems [75]. Meanwhile, conventional numerical methods, such as DP, suffer from a large computational burden and may lead to suboptimal solutions [21]. An example of a comparison between GA and DP is given in [76]. Both methods are used for determining the optimal feed rate profile of a fed-batch culture. The result shows that the final production of monoclonal antibodies (MAb) produced by using a GA is about 24% higher than that produced by using the DP. In addition to the advantage of global solution, GAs can be applied to both "white box" and "black box" models (eg. neural network models) [45, 77]. This offers a great opportunity to combine GAs with neural networks for optimization of fermentation processes.

Finally, AI approaches provide the benefit of rapid prototype development and cost-effective solutions. Due to less *a priori* knowledge being required in AI methods, monitoring, modelling and optimization of fermentation processes can be achieved using a much shorter time as compared to conventional approaches. This can lead to a significant saving in the amount of investment in process development.

1.5 Why is Experimental Investigation Important for Fermentation Study?

Due to practical difficulties and commercial restrictions, many researches [73, 78, 40, 20] have relied only on simulated data based on kinetic or reactor models. However, as mentioned in the context, mathematical models have many limitations. Since the inherent nonlinear dynamics of fermentation processes can not be fully predicted, the process-model mismatching problem could affect the accuracy and applicability of the proposed methodologies.

On the other hand, due to intensive data-driven nature of neural network approaches, a workable neural network model should be trained to adapt to the real environment and should be able to extract the underlying sophisticate relationships between input and output data collected in the experiments. Thus, experimental verification and modification are essential if practical and reliable neural models are required.

1.6 Contributions of the Book

The main contributions of the book are:

- A new neural softsensor is proposed for on-line biomass prediction requiring only the value of DO, feed rate and volume to be measured.
- A novel cascade neural model is developed for modelling the fed-batch fermentation processes. It provides a reliable and efficient representation of the system to be modelled for optimization purposes.
- A new cost-effective methodology, which combine GAs and dynamic neural networks, is established to successfully model and optimize the fed-batch fermentation processes without *a priori* knowledge and detailed kinetics models.
- A new strategy for on-line identification and optimization of fed-batch fermentation processes is proposed using GAs.
- Modified GAs are presented to achieve fast convergence rates as well as global solutions.
- A comparison of a GA and DP has shown that the GA is more powerful for solving high order nonlinear dynamic constrained optimization problems.

1.7 Book Organization

This book consists of eight chapters. Chapter 2 demonstrates the optimization of a fed-batch culture of hybridoma cells using a GA. The optimal feed rate profiles for single feed stream and multiple feed streams are determined via the real-valued GA. The results are compared with the optimal constant feed rate profile. The effect of different subdivision number of the feed rate on the final product is also investigated. Moreover, a comparison between the GA and DP method is made to provide evidence that the GA is more powerful for solving global optimization problems of complex bioprocesses.

Chapter 3 covers the on-line identification and optimization for a high productivity fed-batch culture of hybridoma cells. A series of GAs are employed to identify the fermentation's parameters for a seventh-order nonlinear model and to optimize the feed rate profile. The on-line procedure is divided into three stages: Firstly, a GA is used for identifying the unknown parameters of the model. Secondly, the best feed rate control profiles of glucose and glutamine are found using a GA based on the estimated parameters. Finally, the bioreactor was driven under control of the optimal feed flow rates. The results are compared to those obtained whereby all the parameters are assumed to be known. This chapter shows how GAs can be used to cope with the variation of model parameters from batch to batch.

Chapter 4 develops an on-line neural softsensor for detecting biomass concentration, which is one of the key state variables used in the control and optimization of bioprocesses. This chapter assesses the suitability of using RNNs for on-line biomass estimation in fed-batch fermentation processes. The proposed neural network sensor only requires the DO, feed rate and volume to be measured. Based on a simulated fermentation model, the neural network topology was selected. The prediction ability of the proposed softsensor is further investigated by applying it to a laboratory fermentor. The experimental results are presented, and how the feedback delays affect the prediction accuracy is discussed.

Chapter 5 is devoted to the modelling and optimization of a fed-batch fermentation system using a cascade RNN model and a modified GA. The complex nonlinear relationship between manipulated feed rate and biomass product is described by cascading two softsensors developed in Chapter 4. The feasibility of the proposed neural network model is tested through the optimization procedure using the modified GA, which provides a mechanism to smooth feed rate profiles, whilst the optimal property is still maintained. The optimal feeding trajectories obtained based both on the mechanistic model and the neural network model, and their corresponding yields, are compared to reveal the competence of the proposed neural model.

Chapter 6 details the experimental investigation of the proposed cascade dynamic neural network model by a bench-scale fed-batch fermentation of *Saccharomyces cerevisiae*. A small database is built by collecting data from nine experiments with different feed rate profiles. For a comparison, two neural models and one kinetic model are presented to capture the dynamics of the fed-batch culture. The neural network models are identified through the training and cross validation, while the kinetic model is identified using a GA. Data processing methods are used to improve the robustness of the dynamic neural network model to achieve a closer representation of the process in the presence of varying feed rates. The experimental procedure is also highlighted in this chapter.

Chapter 7 presents the design and implementation of optimal control of fed-batch fermentation processes using a GA based on cascade dynamic neural models and the kinetic model. To achieve fast convergence as well as a global solution, novel constraint handling and incremental feed rate subdivision techniques are proposed. The results of experiments based on different process models are compared, and an intensive discussion on error, convergence and running time are also given.

The general conclusions and thoughts for future research in the area of intelligent biotechnological process control are presented in Chapter 8.

Optimization of Fed-batch Culture of Hybridoma Cells using Genetic Algorithms

Optimizing a fed-batch fermentation of hybridoma cells using a GA is described in this chapter. Optimal single- and multi-feed rate trajectories are determined via the GA to maximize the final production of MAb. The results show that the optimal, varying, feed rate trajectories can significantly improve the final MAb concentration as compared to the optimal constant feed rate trajectory. Moreover, in comparison with DP, the GA- calculated feed trajectories yield a much higher level of MAb concentrations.

2.1 Introduction

Fed-batch processes are of great importance to biochemical industries. Although they typically produce low-volume, high-value products, however, the associated cost is very high. Optimal operation is thus extremely important, since every improvement in the process may result in a significant increase in production yield and saving in production cost. The major objective of the research that is described in this chapter is not to keep the system at a constant set point but to find an optimal control profile to maximize the product of interest at the end of the fed-batch culture. In this work, real-valued GAs are chosen to optimize the high order, dynamic and nonlinear system.

GAs are stochastic global search methods that imitates the principles of natural biological evolution [65, 64, 67, 60]. It evaluates many points in parallel in the parameter space. Hence, it is more likely to converge towards a global solution. It does not assume that the search space is differentiable or continuous and can be also iterated many times on each data received. GAs are a promising and often superior alternative for solving modelling and optimal control problems when conventional search techniques are difficult to use because of severe nonlinearities and discontinuities [79, 76]. Some researches on bioprocess optimization using GAs are found in the literature [80, 81, 76].

GAs operate on populations of strings, which are coded to represent some underlying parameter set. Three operators, selection, crossover and mutation,

L. Z. Chen et al.: *Modelling and Optimization of Biotechnological Processes*, Studies in Computational Intelligence (SCI) **15**, 17–27 (2006)
www.springerlink.com

are applied to the strings to produce new successive strings, which represent a better solution to the problem. These operators are simple, involving nothing more complex than string copying, partial string exchange and random number generation. GA realize an innovative notion exchange among strings and thus connect to our own ideas of human search or discovery.

The remaining sections of this chapter proceed as follows: in Section 2.2, a seventh order model is introduced and the related practical problems are formulated; Section 2.3 explains the basics of GAs; in Section 2.4, the simulation results are given; conclusions are drawn in Section 2.5.

2.2 Proposed Model and Problem Formulation

A seventh order nonlinear kinetic model for a fed-batch culture of hybridoma cells [24] is used in this work. The mass balance equations of a fed-batch fermentation for a single-feed case are:

$$
\begin{aligned}
\frac{dX_v}{dt} &= (\mu - k_d)X_v - \frac{F}{V}X_v \\
\frac{dGlc}{dt} &= (Glc_{in} - Glc)\frac{F}{V} - q_{glc}X_v \\
\frac{dGln}{dt} &= (Gln_{in} - Gln)\frac{F}{V} - q_{gln}X_v \\
\frac{dLac}{dt} &= q_{lac}X_v - \frac{F}{V}Lac \\
\frac{dAmm}{dt} &= q_{amm}X_v - \frac{F}{V}Amm \\
\frac{dMAb}{dt} &= q_{MAb}X_v - \frac{F}{V}MAb \\
\frac{dV}{dt} &= F
\end{aligned}
\tag{2.1}
$$

where, X_v, Glc, Gln, Lac, Amm and MAb are respectively the concentrations in viable cells, glucose, glutamine, lactate, ammonia and MAb; V is the fermentor volume and F the volumetric feed rate; Glc_{in} and Gln_{in} are the concentrations of glucose and glutamine in the feed stream, respectively; Both glucose and glutamine concentrations are used to describe the specific growth rate, μ. The cell death rate, k_d, is governed by lactate, ammonia and glutamine concentrations. The specific MAb production rate, q_{MAb}, is estimated using a variable yield coefficient model related to the physiological state of the culture through the specific growth rate. The parameter values and detailed kinetic expressions for the specific rates, q_{glc}, q_{gln}, q_{lac}, q_{amm} and q_{MAb} are presented in Appendix A.

The multi-feed case which involves two separate feeds F_1 and F_2 for glucose and glutamine respectively is reformulated as follows:

$$
\begin{aligned}
\frac{dX_v}{dt} &= (\mu - k_d)X_v - \frac{F_1+F_2}{V}X_v \\
\frac{dGlc}{dt} &= \frac{F_1}{V}Glc_{in} - \frac{F_1+F_2}{V}Glc - q_{glc}X_v \\
\frac{dGln}{dt} &= \frac{F_2}{V}Gln_{in} - \frac{F_1+F_2}{V}Gln - q_{gln}X_v \\
\frac{dLac}{dt} &= q_{lac}X_v - \frac{F_1+F_2}{V}Lac \\
\frac{dAmm}{dt} &= q_{amm}X_v - \frac{F_1+F_2}{V}Amm \\
\frac{dMAb}{dt} &= q_{MAb}X_v - \frac{F_1+F_2}{V}MAb \\
\frac{dV}{dt} &= F_1 + F_2
\end{aligned}
\tag{2.2}
$$

The criterion to be maximized is the total amount of MAb concentration obtained at the end of the fed-batch culture:

$$J(t_0, t_f) = \max_{F(t)}[MAb(t_f) \cdot V(t_f)] \tag{2.3}$$

The constraints on the control variable and the culture volume are:

$$\begin{array}{rcl} 0 & \leq & F \leq 0.5 \ \ L/day \\ V(t_f) & \leq & 2.0 \ \ L \end{array} \tag{2.4}$$

The following initial culture conditions and feed concentrations are used:

$$\begin{array}{rl} X_v(0) & = 2.0 \times 10^8 \ \ cells/L \\ Glc(0) & = 25 \ \ mM \\ Gln(0) & = 4 \ \ mM \\ Lac(0) & = Amm(0) = MAb(0) = 0 \\ Clc_{in} & = 25 \ \ mM \\ Gln_{in} & = 4 \ \ mM \\ V(0) & = 0.79 \ \ L \end{array} \tag{2.5}$$

2.3 Genetic Algorithm

GAs operate simultaneously on a population of potential solutions applying the principle of natural evolution, i.e., survival of the fittest, to produce better and better approximations to a solution of a problem. At each generation, a new set of approximations (population) is created by the process of selecting individuals according to their level of fitness in the problem domain and breeding them together using operators borrowed from natural genetics. This process leads to the evolution of populations of individuals which are better suited to their environment than the individuals that they are created from. The major elements of the GA operations are:

- *Initialization*, which is usually achieved by generating the required number of individuals using a random number generator. A chromosome representation is needed to describe each individual in the population of interest. The binary representation which is most commonly used in GAs, however, does not yield satisfactory results when applied to multi-dimensional, high precision numerical problems. A more natural representation, real-valued representation, is more attractive for numerical function optimization over binary encodings. With this kind of representation, the computational speed of computers increases as there is no need to convert bit-strings to real values and vice versa, and less memory is required as the floating-point computers can deal with real values directly.
- *Evaluation*, which is done by evaluating the predefined fitness functions. The fitness function is used to provide a measure of how individuals have

performed in the problem domain. In many cases, the fitness function value corresponds to the number of offspring which an individual can expect to produce in the next generation. It is the driving force behind GAs.

- *Selection*, which is performed based upon the individual's fitness such that the better individuals have an increased chance of being selected. There are several schemes for the selection process: roulette wheel selection, scaling techniques, tournament, elitist models, and ranking methods. The first selection method is adopted in this research.

- *Cross-over and mutation*, which are the basic search mechanisms for producing new solutions based on existing solutions in the population. These operators enable the evolutionary process to move towards "promising" regions of the search space. Like their counterpart in nature, crossover produces new individuals which recombine some parts of both parents' genetic material while mutation alters one individual to produce a single new solution. The crossover operation is applied with a probability P_x ("crossover probability" or "crossover rate") when the pairs are chosen for breeding. A mutation operator is introduced to prevent premature convergence to local optima by randomly sampling new points in the search domain and is applied with low probability P_m ("mutation rate").

- *Termination*, which is the end of a run of a GA. A common practice is to terminate a GA after a pre-specified number of generations and then test the quality of the best member of the population against the problem definition. If no acceptable solutions are found, a GA may be restarted or initialized to a fresh search.

In this study, the values of the rate of selection, crossover, and mutation were chosen as 0.08, 0.6, and 0.05 respectively.

2.4 Optimization using Genetic Algorithms based on the Process Model

A simple illustration of optimization using a GA based on the process model is shown in Figure 2.1. The GA generates a control profile $u(t)$, and receives the responses, $\hat{y}(t)$, of the model. According to the cost function J, The GA can eventually find the optimal control profile $u^*(t)$.

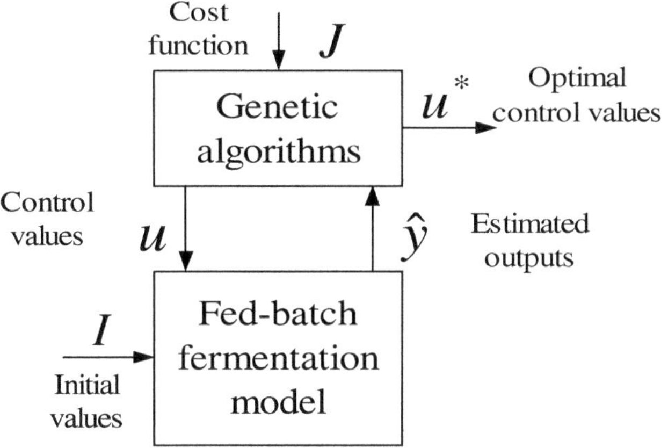

Fig. 2.1. Schematic diagram of optimization using a GA.

2.5 Numerical Results

The GA described previously is used to search for best feeding polices for the system outlined in Section 2.2. The final culture time and volume were fixed to be 10 days and two liters, respectively. For the above problem, suppose

$$p(g) = [p_1(g) \quad p_2(g) \quad \cdots \quad p_q(g)]^T \tag{2.6}$$

is the feed rate matrix for the reactor,

$$p_i(g) = [p_{i1}(g) \quad p_{i2}(g) \quad \cdots \quad p_{im}(g)] \quad (i = 1, 2, \cdots, q) \tag{2.7}$$

is the feed rate vector, where g is the sequence number of a generation, q is the number of the individuals of a generation, and m is the number of intervals within 10 days. The matrix of p forms a population in the GA.

Each individual feed rate vector is constrained by the following conditions:

$$\begin{array}{l} 0 \le \quad p_{ij}(g) \quad \le 0.5 \ L/day \\ 2 \ge \frac{10}{m} \sum_{j=1}^{m} p_{ij}(g) \ (i = 1, 2, \cdots, q; \quad j = 1, 2, \cdots, m) \end{array} \tag{2.8}$$

The performance is measured by the index given as follows:

$$J(0, 10) = MAb(10) \cdot V(10) \tag{2.9}$$

Then the application of the GA to search for the best feed rate profile can be described as follows:

(1) An initial population of feed rate matrix $P(0) = [p_1(0) \quad p_2(0) \quad \cdots \quad p_q(0)]$ is formed with randomly selected individuals.

(2) Each individual $p_1(g)$ is used to calculate the performance index $J(0, 10)$ by solving the non-linear differential equation of the kinetic models. To have the final volume $V(10) = 2\ L$, the initial volume is chosen to be $V(0) = 2 - \frac{10}{m} \sum_{j=1}^{m} p_{ij}(g)$. If $V(0) < 0$, we set $J(0, 10) = 0$.

(3) A new generation $p(g + 1)$, with the same individual number of $p(g)$ is formed by means of reproduction, crossover and mutation based on $p(g)$.

(4) The process will stop if $g = $ maximum generation number. Otherwise, $g = g + 1$, go back to Step 2 to continue.

Single-feed case

The number of intervals within 10 days and the number of individuals of a generation were selected respectively, as $m = 10$ and $q = 1000$. Figure 2.2 shows the best feed rate trajectory, Figure 2.3 and Figure 2.4 present the corresponding histories of the culture volume and the concentration of MAb, respectively. Figure 2.2 shows that the reactor operates as a batch culture for two days before operating in a fed-batch mode. After the batch period of two days, the feeding pattern steadily decreased. This optimal feed rate pattern yielded the final MAb concentration of 155.28 mg/L. The final MAb concentration yielded by the optimal constant feed rate was found to be 141.1 mg/L and its corresponding optimal constant feed rate was 0.136 L/day. Comparing the final MAb concentrations obtained by the optimal varying feed rate and the optimal constant feed rate, the optimal varying feed rate trajectory improved the final MAb concentration by 10%. The time required for the search of the optimal varying feed rate trajectory was about 30 min on a Pentium 100 using MATLAB Genetic Algorithms for Optimization Toolbox (GAOT).

Table 2.1 shows the effect of m on the final level of MAb. It appears that the larger the number of intervals, the higher the final MAb value. However, the larger the number of intervals, the longer the computation time per generation. The computation time required for $m = 20$ is about two times the time required for $m = 10$.

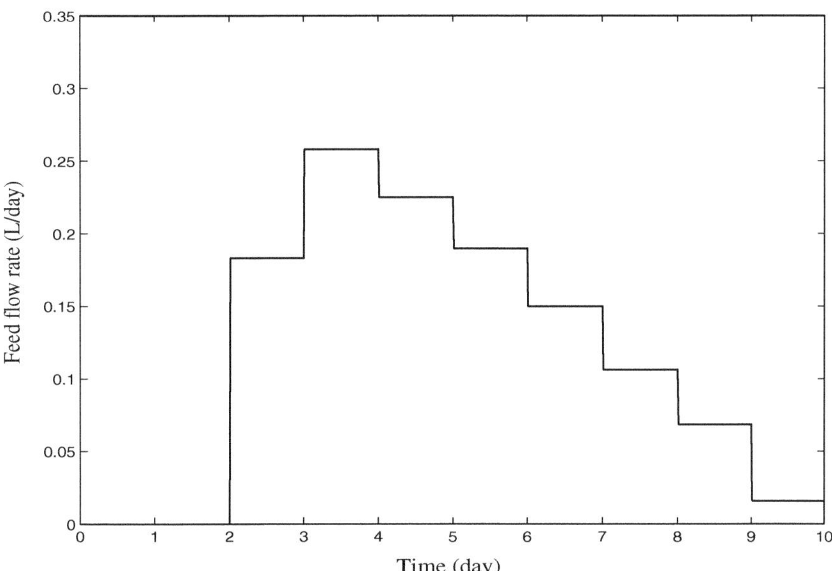

Fig. 2.2. The optimal single-feed rate profile ($m = 10$).

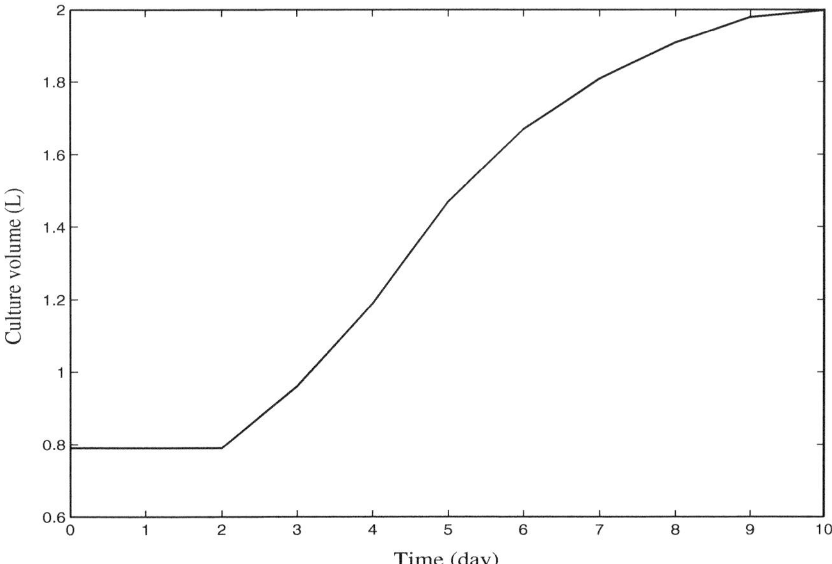

Fig. 2.3. The change of culture volume under control of the optimal single-feed rate profile ($m = 10$).

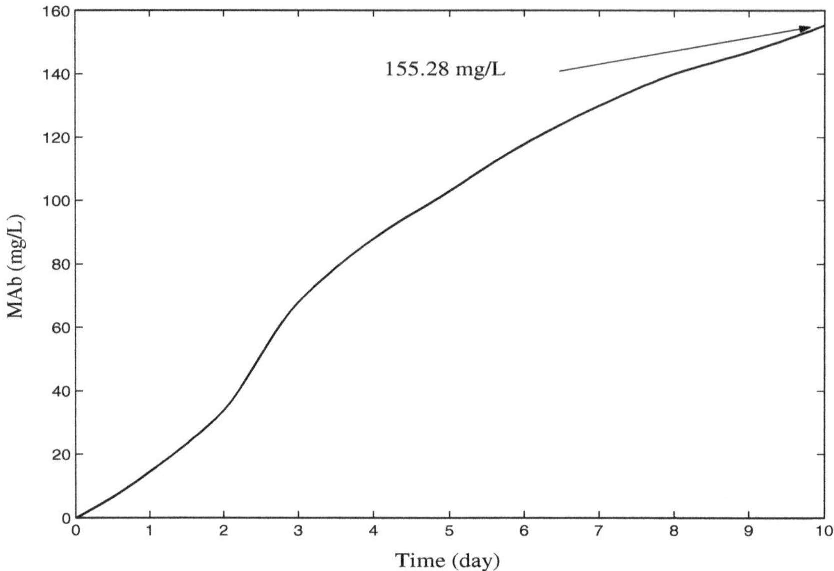

Fig. 2.4. The production of MAb under control of the optimal single-feed rate profile ($m = 10$).

Table 2.1. Effect of m on the final level of MAb (single-feed case).

m	Final level of MAb (mg/L)
5	153.91
10	155.28
15	155.95
20	156.36

Multi-feed case

In the case of multiple feeds, the best trajectories were determined for two separate feeds of glucose and glutamine. Again the number of intervals within 10 days and the number of the individuals of a generation were respectively selected as 10 and 1000 (i.e., $m = 10$ and $q = 1000$). Note that the feed rate vector for this case is:

$$p_i(g) = [p_{i1}(g) \quad p_{i2}(g) \quad \cdots \quad p_{i2m}(g)] \quad (i = 1, 2, \cdots, q) \qquad (2.10)$$

The feed rate vector of the glucose is the first m elements and the feed rate vector of glutamine is the last m elements. Each individual feed rate vector is constrained by the following conditions:

$$0 \leq \quad p_{ij}(g) + p_{i(j+m)}(g) \quad \leq 0.5 \ L/day$$
$$2 \geq \tfrac{10}{m} \sum_{j=1}^{m} [p_{ij}(g) + p_{i(j+m)}(g)] \ (i = 1, 2, \cdots, q; \quad j = 1, 2, \cdots, m) \quad (2.11)$$

Figure 2.5 shows that the glutamine was fed to the reactor first at a rate around 0.251 L/day for five days then followed by a zero rate. On the other hand, the glucose was added after three days at a low rate (0.02 L/day) then followed by a medium rate (0.045 L/day). These trajectories yielded a final MAb concentration of 196.0 mg/L, an improvement of 39% as compared to the optimal constant single-feed rate (0.136 L/day). Figures 2.6 and 2.7 show the corresponding histories of culture volume and MAb, respectively. The determination of the optimal varying feed rate trajectories for the multi-feed case required about three hours on a Pentium 100 using MATLAB GAOT software.

Table 2.2 shows the effect of m on the final level of MAb. It appears that there is not much difference in the final level of MAb for m in this range (five and 20). However, the computation time requires for $m = 20$ is about two times the time required for $m = 10$.

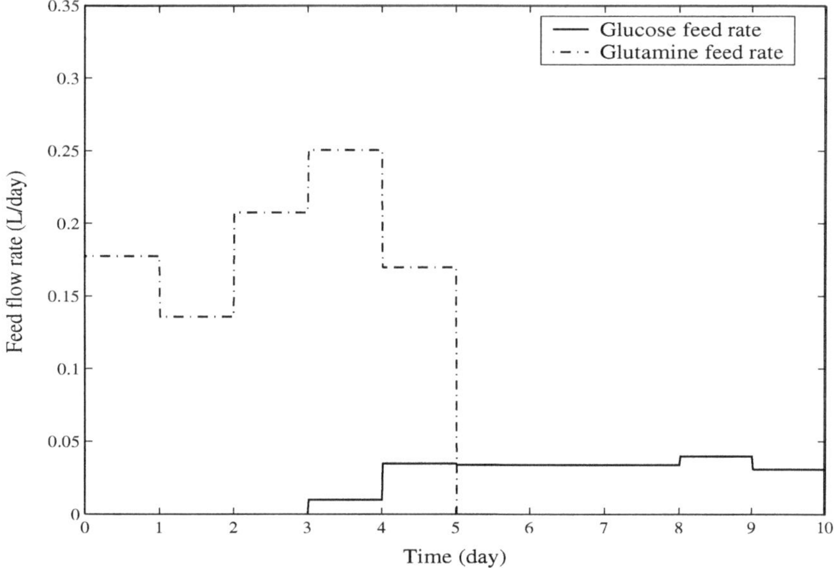

Fig. 2.5. The optimal multi-feed rate profile ($m = 10$).

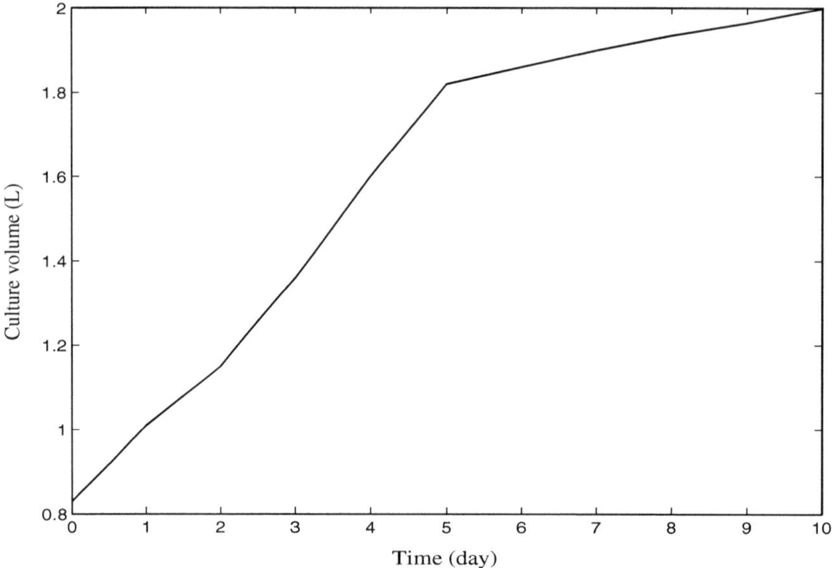

Fig. 2.6. The change of culture volume under control of the optimal multi-feed rate profile ($m = 10$).

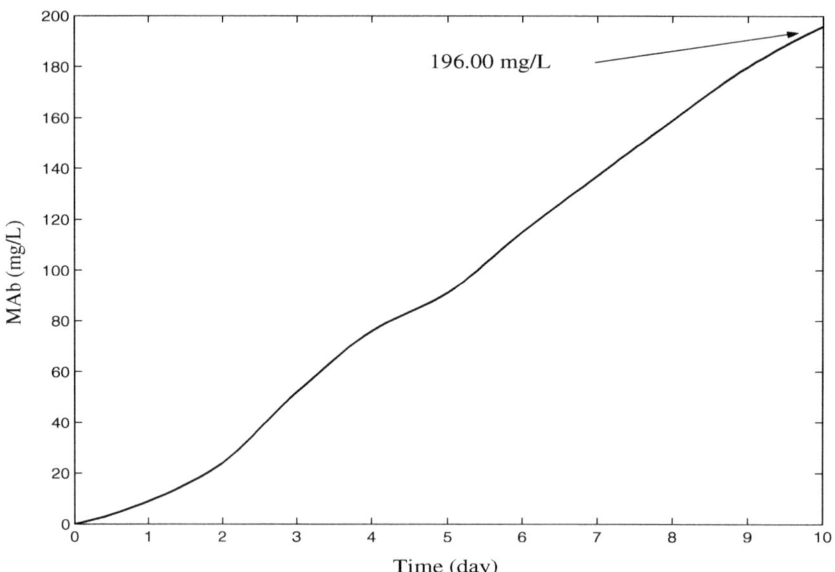

Fig. 2.7. The production of MAb under control of the optimal multi-feed rate profile ($m = 10$).

Table 2.2. Effect of m on the final level of MAb (multi-feed case).

m	Final level of MAb (mg/L)
5	195.05
10	196.00
15	196.16
20	196.27

Table 2.3 shows the comparison between GAs and DP. It is clearly shown that, for all feed rate cases, the GA-calculated feed trajectories yield a higher level of MAb than the DP-calculated feed trajectories.

Table 2.3. Comparison between GA and DP.

Feed rate	MAb (mg/L)
Constant-feed (DP)	134
Constant-feed (GA)	141.1
Single-feed (DP)	147
Single-feed (GA)	156.36
Multi-feed (DP)	158.5
Multi-feed (GA)	196.27

2.6 Conclusions

In this work, the seventh order system is used to describe the fed-batch culture of hybridoma cells and the GA is used to maximize the final MAb production. Optimal feed rate trajectories for a feed stream containing both glucose and glutamine (single-feed case), and separate feed streams of glucose and glutamine (multi-feed case) are searched for using the GA. As compared to the optimal constant feed rate, optimal varying feed rate trajectories are shown to improve the final MAb concentration by 10% for the single-feed rate case and by 39% for the multi-feed rate case. In comparison with the DP method, the GA-calculated feed trajectories increase the final MAb by 5%, 6% and 24% for constant-, single- and multi-feed case, respectively.

3

On-line Identification and Optimization of Feed Rate Profiles for Fed-batch Culture of Hybridoma Cells

This chapter presents an on-line approach for identifying and optimizing fed-batch fermentation processes based on a series of real-valued GA. The model parameters are determined through on-line tuning. The final MAb concentration reaches 98% of the highest MAb concentration obtained in Chapter 2, wherein all the parameters are assumed to be known (i.e., no online tuning). The on-line method proved to be effective in coping with the problem of parameter variation from batch to batch.

3.1 Introduction

The problem of system parameter identification and optimization of control profiles has attracted considerable attention, mostly because of a large number of applications in diverse fields like chemical processes and biomedical systems [22, 23, 25, 82, 83, 79, 81]. To optimize a fed-batch culture, it is essential to have a model, usually a mathematical model, that adequately describes the production kinetics. Based on the mathematical model, the optimal control profiles can be determined to drive the bioreactor to reach the goal. Several techniques, such as the GA described in Chapter 2, have been proposed to determine the optimal control profiles [6, 84, 45, 21, 75].

Practically, the parameters of fermentation models vary from one culture to another. On-line tuning is thus necessary to find accurate and proper values of model parameters and to reduce the process-model mismatching. In this chapter, we intend to use the GA [65, 64, 67, 60] for: i) on-line identifying the parameters of a seventh-order nonlinear model of fed-batch culture of hybridoma cells, and ii) determining the optimal feed rate control profiles for separate feed streams of glucose and glutamine. Finally, we use these control profiles to drive the fermentation process to yield the highest productivity. The salient feature of the approach proposed in this chapter is the on-line model identification, which makes the method more attractive for practical use.

L. Z. Chen et al.: *Modelling and Optimization of Biotechnological Processes*, Studies in Computational Intelligence (SCI) **15**, 29–40 (2006)
www.springerlink.com

The structure of this chapter is as follows: In Section 3.2, a mathematical model describing the kinetics of hybridoma cells [24] is introduced and the related aspects are briefly summarized. The problem of interest is also formulated here. Section 3.3 addresses the methodology proposed in this study. Numerical results are given in Section 3.4. Section 3.5 summarizes the work that is presented in this chapter.

3.2 Fed-batch Model and Problem Formulation

A mathematical model for a fed-batch culture of hybridoma cells [24] is employed in this study. The details are given in Appendix A. The mass balance equations for the system in fed-batch mode of a multi-feed case and the problem formulation are presented below.

The multi-feed case, which involves two separate feeds F_1 and F_2 for glucose and glutamine respectively, is reformulated as follows:

$$
\begin{aligned}
\frac{dX_v}{dt} &= (\mu - k_d)X_v - \frac{F_1 + F_2}{V}X_v \\
\frac{dGlc}{dt} &= (Glc_{in} - Glc)\frac{F_1 + F_2}{V} - q_{glc}X_v \\
\frac{dGln}{dt} &= (Gln_{in} - Gln)\frac{F_1 + F_2}{V} - q_{gln}X_v \\
\frac{dLac}{dt} &= q_{lac}X_v - \frac{F_1 + F_2}{V}Lac \\
\frac{dAmm}{dt} &= q_{amm}X_v - \frac{F_1 + F_2}{V}Amm \\
\frac{dMAb}{dt} &= q_{MAb}X_v - \frac{F_1 + F_2}{V}MAb \\
\frac{dV}{dt} &= F_1 + F_2
\end{aligned}
\tag{3.1}
$$

where, X_v, Glc, Gln, Lac, Amm and MAb are respectively the concentrations in viable cells, glucose, glutamine, lactate, ammonia and MAb; V is the fermentor volume and F the volumetric feed rate; Glc_{in} and Gln_{in} are the concentrations of glucose and glutamine in the feed stream, respectively. The parameter values and kinetic expressions are given in Appendix A.

In this work, there are two problems that need to be solved:

(1) The first problem is to estimate all sixteen parameters, μ_{max}, k_{dmax}, $Y_{xv/glc}$, $Y_{xv/gln}$, m_{glc}, k_{mglc}, K_{glc}, K_{gln}, α_0, K_μ, β, k_{dlac}, k_{damm}, k_{dgln}, $Y_{lac/glc}$, and $Y_{amm/gln}$, from the measured values of X_v, Glc, Gln, Lac, Amm, MAb and V at the beginning of the fed-batch fermentation fed with a deliberate single-feed stream. The structure of the kinetic model used for the study is known (as shown in Appendix A). The identification problem is to minimize the error between actual values of these state variables and their estimated values predicted from the estimated parameters. The objective function is as follows:

$$
J_I(t_0, t_N) = \min_P \|I(t_0), I(t_1)\cdots, I(t_N)\|
\tag{3.2}
$$

with

$$I(t_i) = \min_P \|X_{est}(t_i) - X(t_i)\|, \quad i = 1, 2, \cdots, N \tag{3.3}$$

where, P, $X_{est}(t_i)$ and $X(t_i)$ are the estimated parameters, the estimated state variables and the actual (measured) state variables, respectively; $\|\cdot\|$ is the notation for $\mathcal{L}_2$ norm; N is the total number of intervals of the reaction time.

(2) The second problem is to determine how the glucose and glutamine should be fed to the fermentor in order to drive MAb to the maximum, for a set of initial conditions and constraints. The criterion used is the total amount of MAb obtained at the end of the fed-batch fermentation:

$$J_0(t_0, t_f) = \max_{F(t)} [MAb(t_f) \cdot V(t_f)] \tag{3.4}$$

The constraints on the control variable and the culture volume are:

$$\begin{aligned} 0 &\leq F \leq 0.5L/d \\ V(t_f) &\leq 2.0L \end{aligned} \tag{3.5}$$

The following initial culture conditions and feed concentrations have been used:

$$\begin{aligned} X_v(0) &= 2.0 \times 10^8 cells/L \\ Glc(0) &= 25mM \\ Gln(0) &= 4mM \\ Lac(0) &= Amm(0) = MAb(0) = 0 \\ Clc_{in} &= 25mM \\ Gln_{in} &= 4mM \\ V(0) &= 0.79L \end{aligned} \tag{3.6}$$

The above mathematical models and initial conditions have been used to generate a 'reality' for testing the schemes proposed in the study.

3.3 Methodology Proposed

The methodology proposed for on-line operation is composed of three steps as shown in Figure 3.1: Step 1: On-line identification of system parameters; Step 2: Optimization of feed rate control profiles; Step 3: Application of the optimal feed rate control profiles.

Step 1 On-line identification of system parameters

A deliberate inlet single feed stream that is fed to the hybridoma cells culture is used to identify the kinetic model. The actual values of the state variables are measured at every sampling time, and the estimated values of state variables are calculated from the model based on the candidate solutions (parameters) of a real-valued GA at the same time. Both measured values and estimated values of state variables are used to evaluate the fitness of individuals using the objective function defined in Equation 3.3. At the termination of the GA for each sampling data, the best population is stored. All the best populations obtained are added together as an initial population which carry the information of the parameters for the whole sampling period instead of one particular sampling point. The GA is run again using the initial population and the objective function, which is described in Equation 3.2. The aim is to minimize the error between actual values and estimated values for the whole samples instead of one sample. This *chosen initial population* and *variation in objective function* may prevent the GA from premature convergence which will lead the GA stuck in a local minimum.

Step 2 Optimization of feed rate control profiles

In this step, the optimal multi-feeding control profiles are worked out based on the estimated model obtained from the previous step. The time axis of the control trajectories (from the end of identification to the end of fermentation) is discretized into a number of steps. The control values at each step are the variables to be optimized by the GA and become the elements of the chromosomes. The GA creates candidate solutions in the form of floating-point representation of variables: chromosomes. These candidate solutions are real values with random numbers within the search domain; ie. constraints on the feed flow (e.g. $0 \leq F \leq 0.5L/d$). A numerical integration method is used to simulate the system for each chromosome in the population. Subsequently, the resulting objective values for the different chromosomes are evaluated and used for the selection. The program is stopped when a predefined maximum number of iteration is reached. Constraints on state variables (e.g. maximum volume) are implemented by penalties in the objective function.

Step 3 Application of the optimal control profiles

In this step, the fed-batch culture of hybridoma cells is run automatically under the control of optimal feed rate control profiles obtained from Step 2.

3.4 Numerical Results

The identification and optimization procedure described previously was used to estimate the parameters of the system kinetic expression and to determine the best utilization of a given volume of culture medium in order to maximize the productivity of a hybridoma cells culture. The total fermentation time was 10 days including both identification and the optimal control period. The final culture volume was fixed to be $2L$.

In this study, the values of the rate of selection, crossover, and mutation in the GA were chosen as 0.08, 0.6, and 0.05 respectively.

Identification of system parameters

The time used for parameter identification was the first two days of the fed-batch fermentation. The initial conditions were given by Equation A.4 in Appendix A. The model equations of the single-feed rate hybridoma cell culture

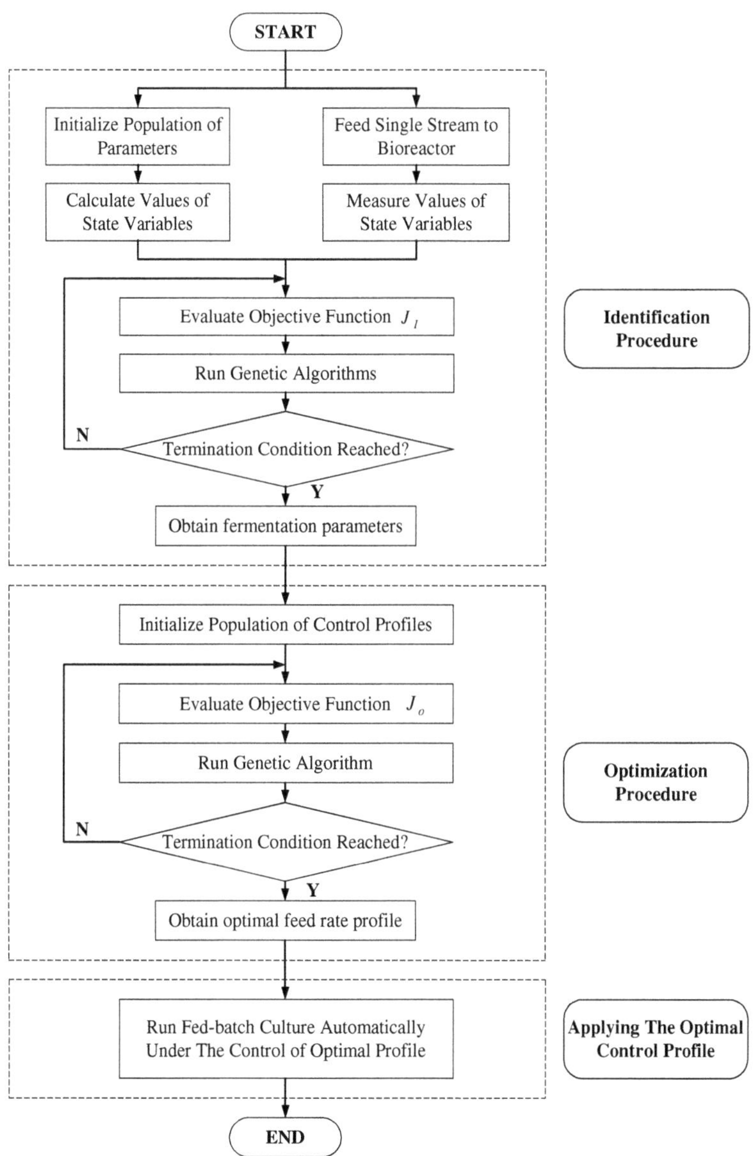

Fig. 3.1. Schematic diagram of the methodology proposed.

were also given in Appendix A. Ten samples measured in equal time length were used as actual state variables. A sampling time of 0.2 day was used. The deliberate inlet single-feed flow rate employed for the two-day identification period was as follows:

$$F = 0.005 - (-1)^n \times 0.0025 \, , \qquad n = 0, 1, \cdots, 10 \qquad (3.7)$$

where, n is the number of samples.

The upper and lower bounds on the variables to be estimated was set to be ± 50 percent of the actual values. The initial population size was 50 for the GA at each sampling point. The GA was run for 50 generations for each measured input-output data pair, and the best population found by the GA at each sample was stored. In order to find a system model which is as close as possible to the actual model instead of suboptimal results, the whole best populations were stored, and were used as an initial population of the GA which was to be run for another 200 generations. The GA took a total of 700 generations to estimate the parameters. The on-line identification procedure is illustrated in Figure 3.2.

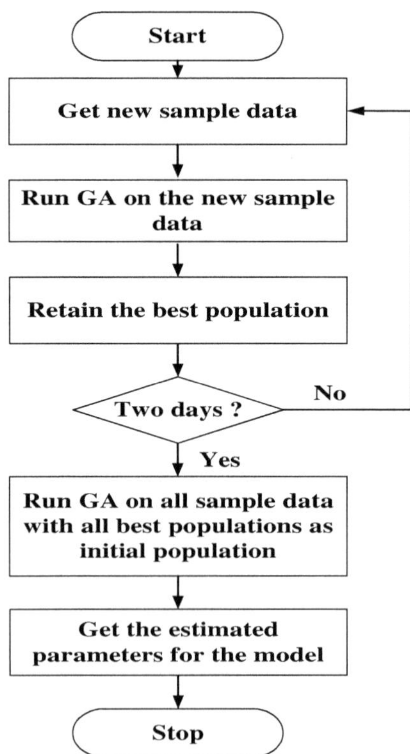

Fig. 3.2. On-line parameter identification procedure using genetic algorithms.

The estimation of parameters is shown in Figure 3.3 and Figure 3.4. The percentage of error is shown in Table 3.1. The time spent for running 50 generations of the GA at each sample was about 30 minutes on a Pentium II Celeron 300MHz computer using MATLAB GAOT software. The time required for the last 200 generations was about 2 hours which is negligible when compared to the long fermentation period (10 days).

From Figure 3.3 and Figure 3.4, one can see that most parameters converge to stable values after 500 generations. Table 3.1 shows that some of the final estimated parameters have very large percentage errors. In the following sections, via simulation, it is shown that these large percentage errors have little effect on the final level of MAb.

Table 3.1. Parameter values obtained by on-line identification.

Parameters	Actual values	Estimated values	Percentage of error (%)
μ_{max}	1.09	1.0848	0.47
$Y_{xv/glc}$	1.09	1.1525	5.7
m_{glc}	0.17	0.2122	24.8
K_{glc}	1.0	0.6394	36.1
α_0	2.57	2.5599	0.4
β	0.35	0.3532	0.9
k_{damm}	0.06	0.0304	49.3
$Y_{lac/glc}$	1.8	1.7998	0.01
k_{dmax}	0.69	0.5418	21.5
$Y_{xv/gln}$	3.8	3.8062	0.16
k_{mglc}	19.0	11.5548	39.2
K_{gln}	0.3	0.3411	13.7
K_{μ}	0.02	0.0176	11.8
k_{dlac}	0.01	0.0106	6.1
k_{dgln}	0.02	0.0210	5.0
$Y_{amm/gln}$	0.85	0.8539	0.5

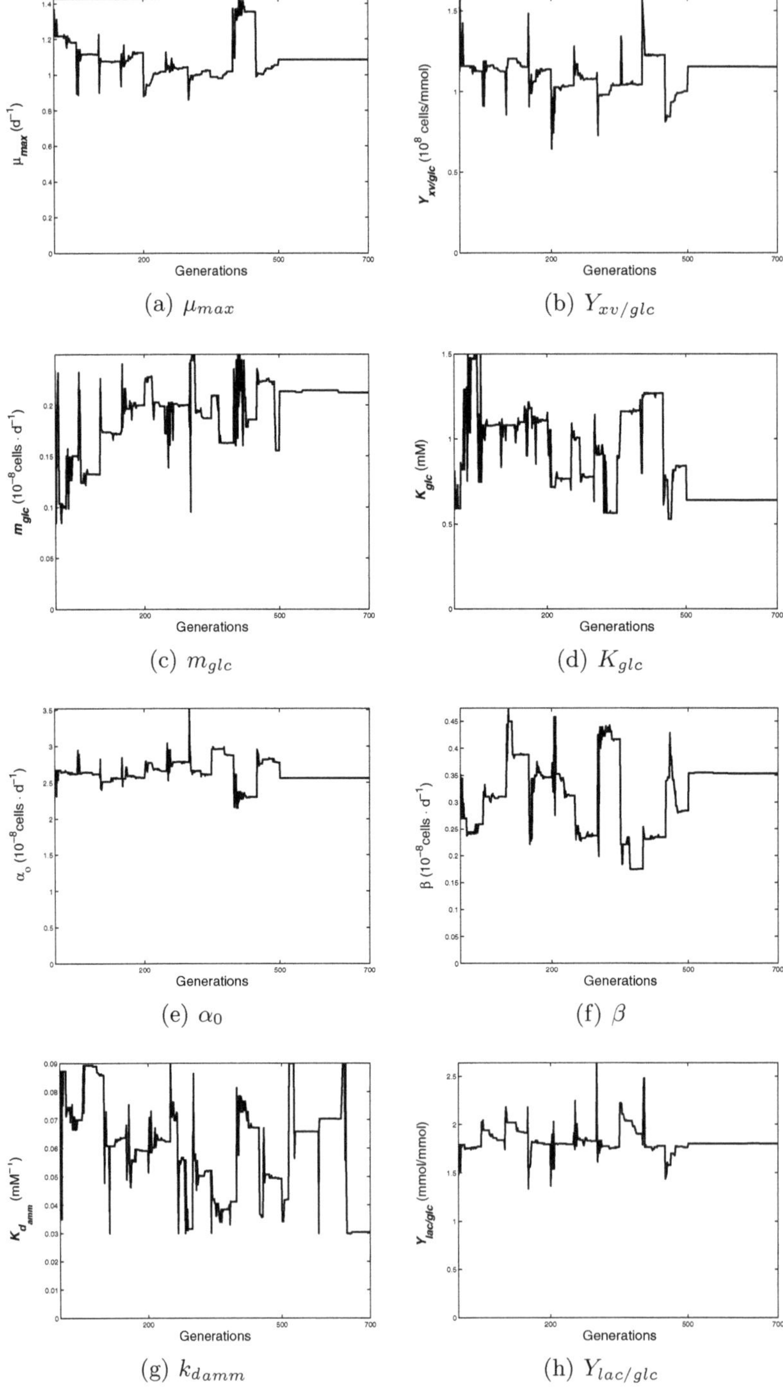

Fig. 3.3. Identification of system parameters.

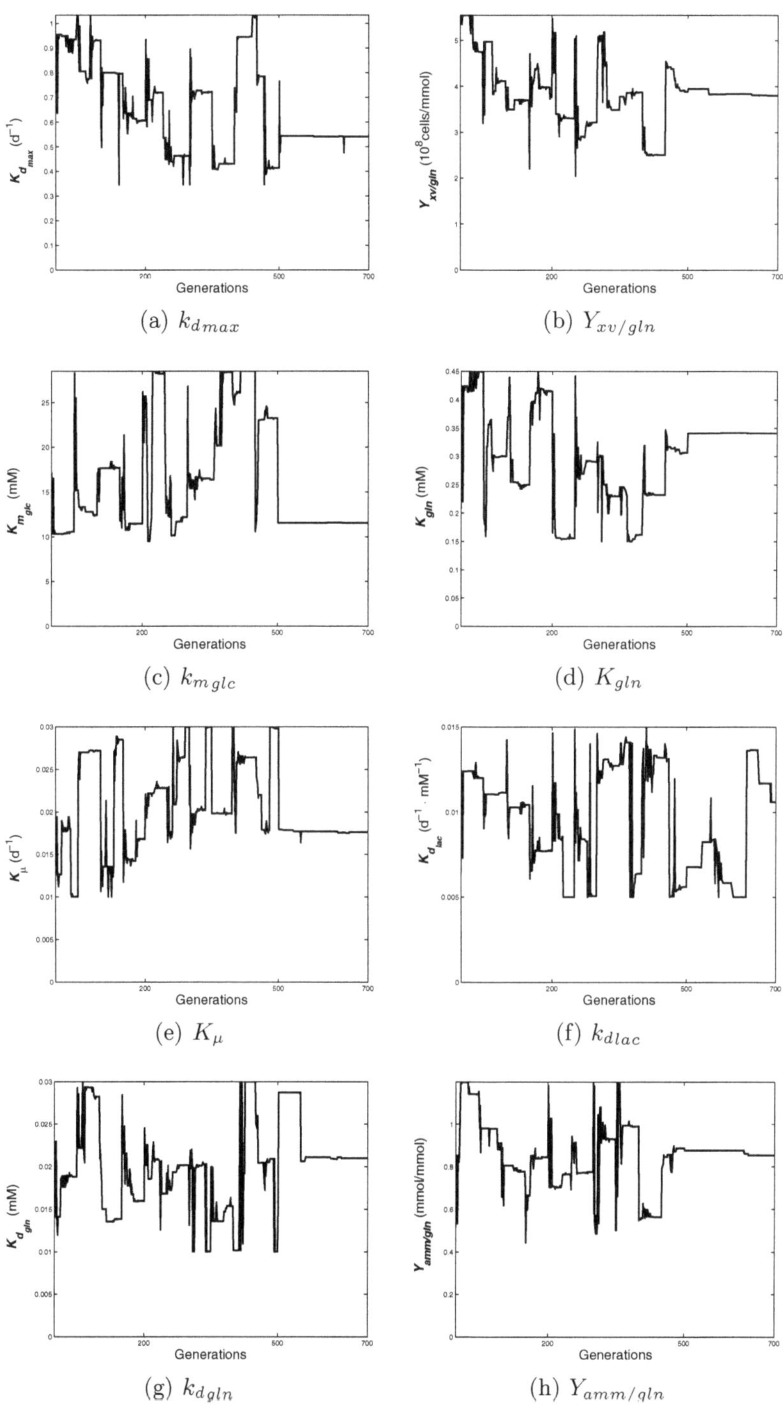

(a) k_{dmax}

(b) $Y_{xv/gln}$

(c) k_{mglc}

(d) K_{gln}

(e) K_{μ}

(f) k_{dlac}

(g) k_{dgln}

(h) $Y_{amm/gln}$

Fig. 3.4. Identification of system parameters (continued).

Optimization of multi-feed rate control profiles

The feed flow rates, F_1 and F_2, were discretized into a set of four piece-wise constant control actions. Each of them has an interval of two days. The initial volume which was the final volume of the identification process was $0.8L$, and the initial population size was chosen to be 200. The optimization procedure using the GA was run for 200 generations. Optimal control profiles which were determined for two separate feeds of glucose and glutamine are shown in Figure 3.5. The optimal feeding rates actually started from day two of the culture. Glutamine was fed to the culture first at a high rate ($0.46\ L/d$) from day two to day four, then followed by a lower rate ($0.027\ L/d$). On the other hand, glucose was added at a lower rate ($0.014\ L/d$) from day two to day four followed by a medium rate ($\sim 0.035 L/d$).

The time required for the optimization step was about two hours which can also be ignored when compared to the 10 days culture time.

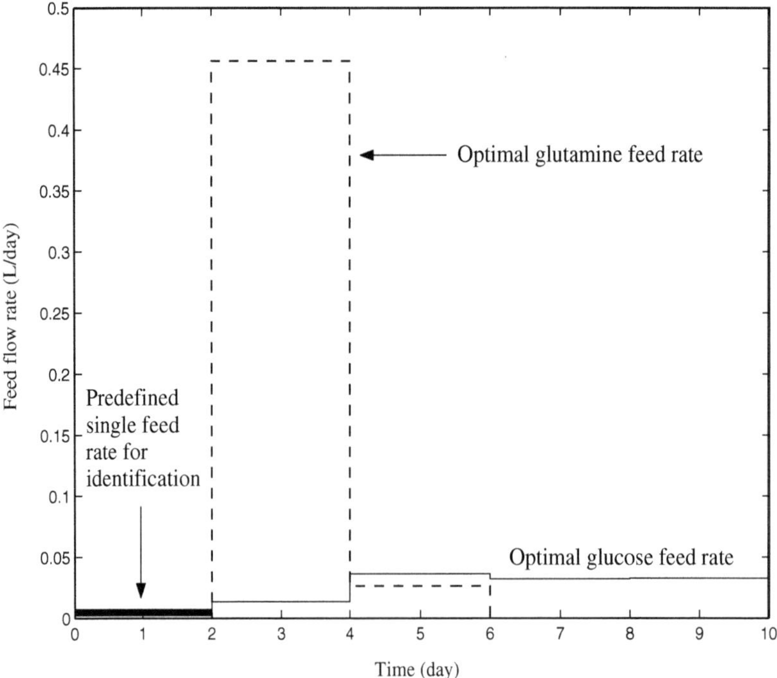

Fig. 3.5. Feed rate profiles for identification (single-feed) and optimization (multi-feed), the single feed rate for identification is defined by Equation 3.7.

Application of the optimal feed flow rates to the bioreactor

The time for optimal control was eight days (the first two days of fermentation were used for parameters estimation). The optimal profiles obtained from the previous stage yielded a final MAb concentration of 193.1 mg/L and a final volume of $2L$ as shown in Figure 3.6 and Figure 3.7 respectively. The performance achieved by this on-line optimization is 2% less than the best result (196.27 mg/L) obtained in the case whereby all the parameters are assumed to be known in Chapter 2. This shows that the proposed real-valued GA approach can be a good alternative method for solving on-line identification and optimization problems.

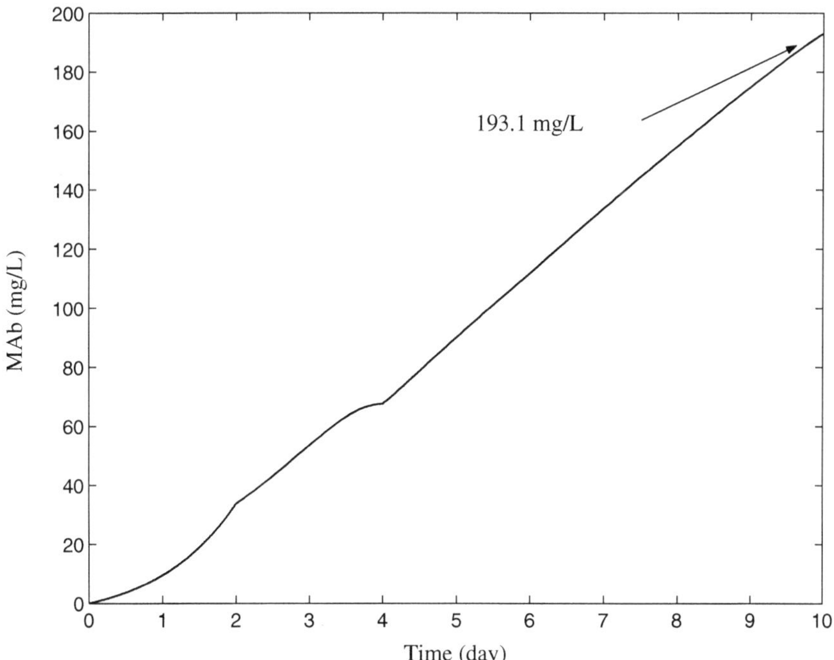

Fig. 3.6. The production of MAb.

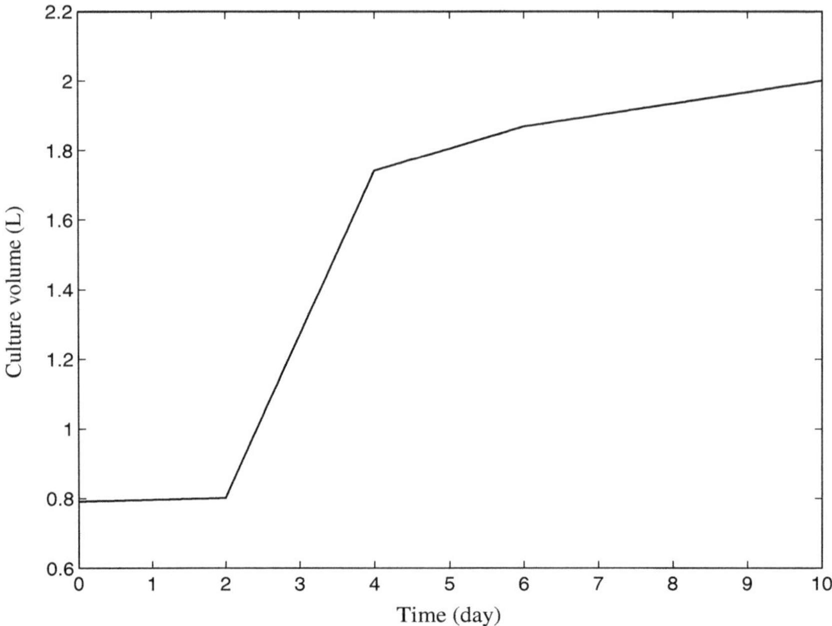

Fig. 3.7. The history of culture volume.

3.5 Summary

In this chapter, an on-line identification and optimization method, based on a series of real-valued GA, is studied for a seventh-order nonlinear model of fed-batch culture of hybridoma cells. The parameters of the model are assumed to be unknown. The on-line procedure is divided into three stages: Firstly, the GA is used for identifying the unknown parameters of the model. Secondly, the best feed rate control profiles of glucose and glutamine are found by the GA based on the estimated parameters. Finally, the bioreactor is driven under the control of the optimal feed flow rates. The final MAb concentration of 193.1 mg/L and a final volume of $2L$ are reached at the end of the fermentation. This result is only 2% less than the best result (196.27 mg/L) obtained for the case wherein all the parameters are assumed to be known (i.e., no online identification). The real-valued GA have proved to be effective tools for solving on-line identification and optimization problems.

4

On-line Softsensor Development
for Biomass Measurements using Dynamic
Neural Networks

One of the difficulties encountered in control and optimization of bioprocesses is the lack of reliable on-line sensors, which can measure the key processes' state variables. This chapter assesses the suitability of using RNNs for on-line biomass estimation in fed-batch fermentation processes. The proposed neural network sensor only requires the DO, feed rate and volume to be measured. The results show that RNNs are a powerful tool for implementing an on line biomass softsensor in experimental fermentations.

4.1 Introduction

At the heart of bioprocess control is the ability to monitor important process variables such as biomass concentration [43]. The lack of reliable on-line sensors, which can accurately detect the important state variables, is one of the major challenges of controlling bioprocess accurately, automatically and optimally in biochemical industries [14, 2, 7, 85]. Softsensors (also called software sensors) have been considered as alternative approaches to this problem [86, 10, 72, 73, 74, 40]. In this chapter, RNNs with both activation feedback and output feedback connections are used for on-line biomass prediction of fed-batch baker's yeast fermentation. The information that is required by the softsensor involves the concentration of DO, feed flow rate and the reaction volume.

Softsensors work in a manner of cause and effect, the inherent biologic relation between measured and unmeasured states could affect the prediction accuracy significantly. DO, pH values, concentrations of carbon dioxide and ethanol are the most commonly selected process variables, which can be readily measured on-line in a research laboratory using standard sensors. Among them, DO concentration, which reflects the fundamental level of energy transduction in bioreaction, is intricately linked to cellular metabolism. It changes about 10 times faster than the cell mass and substrate concentrations during the reaction course. Some researchers showed that controlling DO at or above

L. Z. Chen et al.: *Modelling and Optimization of Biotechnological Processes*, Studies in Computational Intelligence (SCI) **15**, 41–56 (2006)
www.springerlink.com

a critical value could enhance the performance of the bioreactor [87, 88]. Nor et al. studied the on-line application of DO concentration [89]. They estimated the specific growth rate of fed-batch culture of *Kluyveromyces fragilis* based on the measurement of the maximum substrate uptake rate (MSUR) and on-line DO concentration using mass balance equations. The main assumptions made were that the specific growth rate and the cell yield remained constant during each feeding interval and that the culture was carbon-source limited. The information of on-line DO concentration was also employed to detect the acetate formation in *Escherichia coli* cultures [90]. Acetate accumulation in fed-batch cultivations is detrimental to the recombinant protein production. On-line detection of acetate enables the development of feedback control strategies for substrate feeding that avoids acetate accumulation, thus increasing the production of recombinant protein. Therefore, there is no doubt that DO dynamics are strongly related to the environmental conditions, and thus an appropriate process variable for on-line inferential estimation of biomass concentration.

The main features of the proposed on-line softsensor are: i) only the DO concentration, feed rate and volume are required to be measured; ii) RNNs are used for predicting the biomass concentration; iii) Neither *a priori* information, nor a moving window technique is necessary.

The layout of the remainder of the chapter is as follows: in Section 4.2, the recurrent neural softsensor structure is given and the simulation studies are described; in Section 4.3, the experimental investigation is detailed and the results are discussed; conclusions are drawn in Section 4.4.

4.2 Softsensor Structure Determination and Implementation

Recurrent neural network softsensor model

A RNN is chosen to estimate the biomass concentration because of its strong capability of capturing the dynamic information underlying the input-output data pairs. The configuration selection of a RNN is problem specific. In this study, an extended, fully-connected RNN, known as the Williams-Zipser network [37, 91], is used for on-line biomass estimation in the fermentation process due to the "dynamically rich" nature of this kind of network. Selection of a suitable RNN topology is based on simulation data generated by a mathematical model. The suitable RNN topology is then re-trained using experimental data. A fine-tuning of the RNN is necessary to make it adaptable to the real environment.

The structure of the proposed neural softsensor is given in Figure 4.1. The inputs of the neural sensor are feed rate F, volume V and DO, which are all continuously available. The output of the sensor gives the estimated biomass concentration. This neural network consists of TDLs, one hidden layer, one

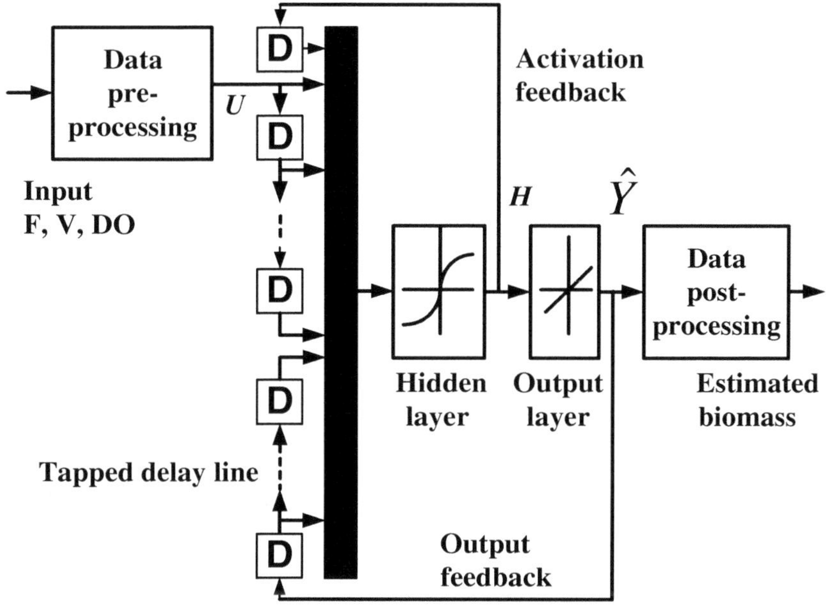

Fig. 4.1. Structure of the proposed recurrent neural softsensor.

output neuron, feed-forward paths and feedback paths. All connections could be multiple paths. In order to enhance dynamic behaviors of the sensor, outputs from the output layer (output feedback) and the hidden layer (activation feedback) are connected to the input layer through TDLs. The output of the i-th neuron in the hidden layer is of the form:

$$h_i(t) = f_h(\sum_{j=0}^{n_a} W_{ij}^I\, p(t-j) + \sum_{k=1}^{n_b} W_{ik}^R\, \hat{y}(t-k) + \cdots$$

$$\cdots + \sum_{l=1}^{n_c} W_{il}^H\, h_l(t-1) + b_i^H) \tag{4.1}$$

where, p is the neural network input, $\hat{y}$ is the neural network output and h is the hidden neuron's output; b_i^H is the bias of i-th hidden neuron; n_a, n_b, n_c are the number of input delays, the number of output feedback delays and the number of hidden neurons, respectively;

$f_h(\cdot)$ is a sigmoidal function;

W_{ij}^I is the weight connecting the j-th delayed input to i-th hidden neuron;

W_{ik}^R is the weight connecting the k-th delayed output feedback to i-th hidden neuron;

W_{il}^H is the weight connecting the l-th hidden neuron output feedback to the i-th hidden neuron.

Only one neuron is placed in the output layer, so the output is:

$$\hat{y}(t) = f_Y \Big(\sum_{m=1}^{n_c} W_m^Y \, h_m(t) + b_Y \Big) \tag{4.2}$$

where, $f_Y(\cdot)$ is a pure linear function;

W_m^Y is the weight connecting the m-th hidden neuron's output to the output neuron;

b_Y is the output neuron bias.

Simulation study

A mathematical model, which is governed by a set of differential equations derived from mass balances in fed-batch fermentation processes [17, 92], was used to generate simulation data. The details of the model are given in Appendix B. A schematic illustration of the simulated fermentation model is shown in Figure 4.2. Three output variables, biomass, DO and volume were generated from a given feed rate by solving the differential equations.

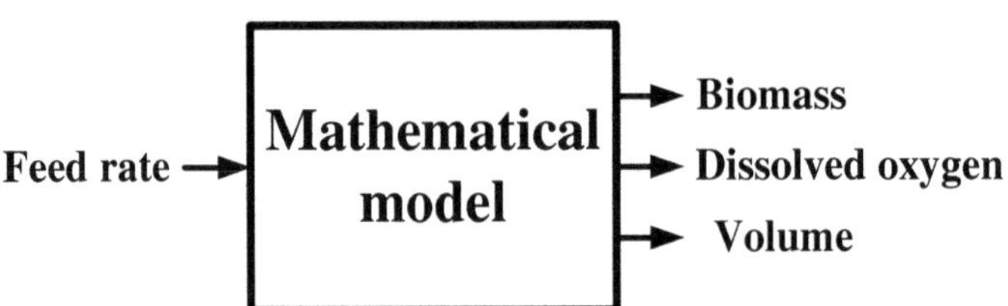

Fig. 4.2. Schematic illustration of the simulated fermentation model.

Five different feed rate profiles were chosen to excite the mathematical fermentation model: (1) a square-wave feed flow, (2) a saw-wave feed flow, (3) a stair-shape feed flow, (4) an industrial-feeding policy and, (5) a random-steps feed flow. These feed rates are shown in Figure 4.3. Each of the first four feed rate profiles yielded 150 input-output (target) pairs corresponding to six minutes sampling time during a 15-hour fermentation; the random-step feed rate yielded 450 data pairs during a 45-hour fermentation with the same length of sampling interval.

A general procedure for developing neural networks [33] are: (1) data preprocessing, (2) appropriate training procedure, (3) generalization and, (4) topology optimization.

Before training a RNN, the input and target data are pre-processed (scaled), thus they are within a specified range, [-1, 1]. This specified range is the most sensitive area of the sigmoidal function, which is the hidden layer activation function. In this case, the output of the trained network will also

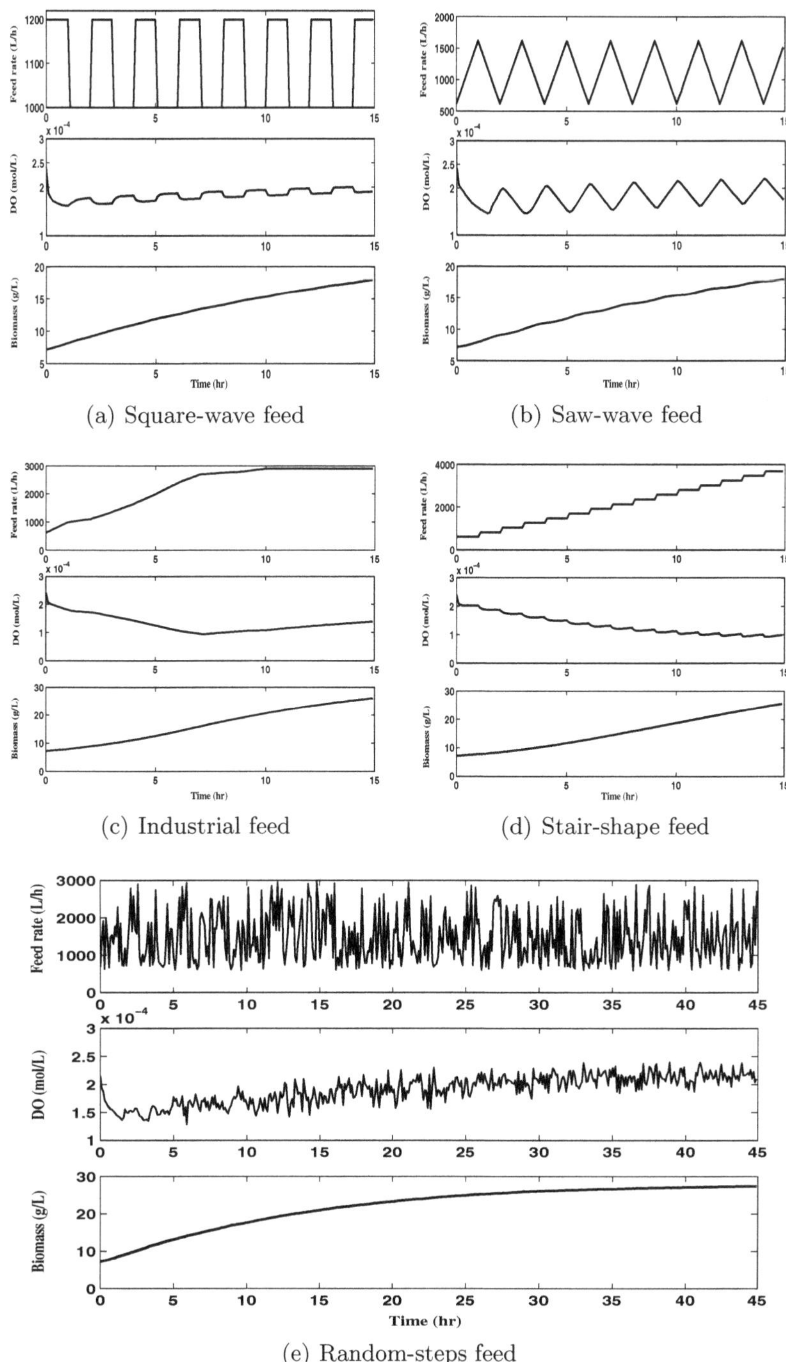

(a) Square-wave feed

(b) Saw-wave feed

(c) Industrial feed

(d) Stair-shape feed

(e) Random-steps feed

Fig. 4.3. Plots of simulation data for five different feed rates.

be in the range [-1,1]. A post-processing procedure has to be performed in order to convert the output back to its original unit.

The performance function that is used for training the neural networks is a mean square error (MSE):

$$MSE = \frac{1}{N}\sum_{i=1}^{N}(X_i^a - \hat{X}_i)^2 \tag{4.3}$$

where, N is the number of training data pairs; X_i^a is the target (actual) value of biomass; $\hat{X}_i$ is the corresponding estimated value preduced by the neural softsensors.

The Levenberg-Marquardt backpropagation (LMBP) training algorithm is adopted to train the neural networks due to its faster convergence and memory efficiency [34, 93]. The algorithm can be summarized as follows:

1. Present input sequence to the network. Compute the corresponding network outputs with respect to the parameters (i.e., weights and bias) $\mathbf{X}_k$. Compute the error $\mathbf{e}$ and the overall MSE error.
2. Calculate the Jacobian matrix $\mathbf{J}$ through the backpropagation of Marquardt sensitivities from the final layer of the network to the first layer.
3. Calculate the step size for updating network parameters using:

$$\Delta\mathbf{X}_k = -[\mathbf{J}^T(\mathbf{x}_k)\mathbf{J}(\mathbf{x}_k) + \mu_k\mathbf{I}]^{-1}\mathbf{J}^T(\mathbf{x}_k)\mathbf{e} \tag{4.4}$$

   where, μ_k is initially chosen as a small positive value (e.g., $\mu_k = 0.01$).
4. Recompute the MSE error using $\mathbf{X}_k + \Delta\mathbf{X}_k$. If this new MSE error is smaller than that computed in step 1, then decrease μ_k, let $\mathbf{X}_{k+1} = \mathbf{X}_k + \Delta\mathbf{X}_k$ and go back to step 1. If the new MSE error is not reduced, then increase μ_k and go back to step 3.

The algorithm terminates when i) the norm of gradient is less than some predetermined value or, ii) MSE error has been reduced to some error goal or, iii) μ_k is too large to be increased practically or, iv) a predefined maximum number of iterations has been reached.

Data generated from the five different feed rate profiles were divided in three groups: the training data set, the validation data set and the testing data set. A well known fact of choosing the training data set is the training data set has to cover the entire state space of the system as many times as possible. In this study, the random-steps, which excited the process the most, was used to generate the training data set. Another set of data generated from the stair-shape feed rate was used as the validation data set. To prevent the neural network from being over-trained, an early stopping method was used. The error on the validation set was monitored during the training process. The validation error would normally decrease during the initial phase of training. However, when the network began to over-fit the data, the error on the validation set would typically begin to rise. When the validation error increased for a specified number of iterations, the training was stopped, and the weights and biases at the minimum of the validation error were obtained. The rest of the data sets, which were not seen by the neural network during the training and validation period, were used in examining the trained network.

There are no general rules or guidelines for selection of the optimal number of hidden neurons in RNNs [39, 78]. The most commonly used method is trial and error. Fewer neurons results in inadequate learning by the network; while too many neurons create over-training and result in poor generalization. One straightforward approach adopted by many researchers is to start with the smallest possible network and gradually increase the size until the performance begins to level off [39, 94, 95, 96]. From an engineering point of view, however, the smallest possible size of a neural network, which can solve the problem, is the desired end result. The approach, which works in the opposite way to the method mentioned above, was used in this work. Starting with a reasonably big network, it was then gradually "shrunk" until the error appearing on the test data was beyond acceptance.

Simulation results

The evaluation function that is used for testing the neural networks is a root mean squared percentage (RMSP) error index [57], which is defined as:

$$E = \sqrt{\frac{\sum_{t=1}^{N}(X_t^m - \hat{X}_t)^2}{\sum_{t=1}^{N}(X_t^m)^2}} \times 100 \tag{4.5}$$

where, N is the number of sampling data pairs; X_t^m is the measured (actual) value of biomass at sampling time t; $\hat{X}_t$ is the corresponding estimated value predicted by the neural softsensors.

The RMSP error between the network output and the measured output of test data set was used to evaluate the merit of the network. In this study, extensive test simulations were carried out. For each network structure, 150 networks were trained; the one that produced the smallest RMSP error for the test data sets was retained. The selection was finally narrowed to two choices: six hidden neurons and 12 neurons. A representative set of error distributions is shown in Figure 4.4 for various combinations of delays and the number of hidden neurons. As shown in the Figure, 12-hidden-neuron networks frequently out-performed the six-hidden neuron networks with the exception of the "0/4/1" structure (zero input delay, four output feedback delays and one activation feedback delay). The testing RMSP errors for these two network structures were very close, and were smaller than other types of network structures. The network with six hidden neurons was therefore chosen for the on-line biomass estimation because of the small prediction error and small size of the network.

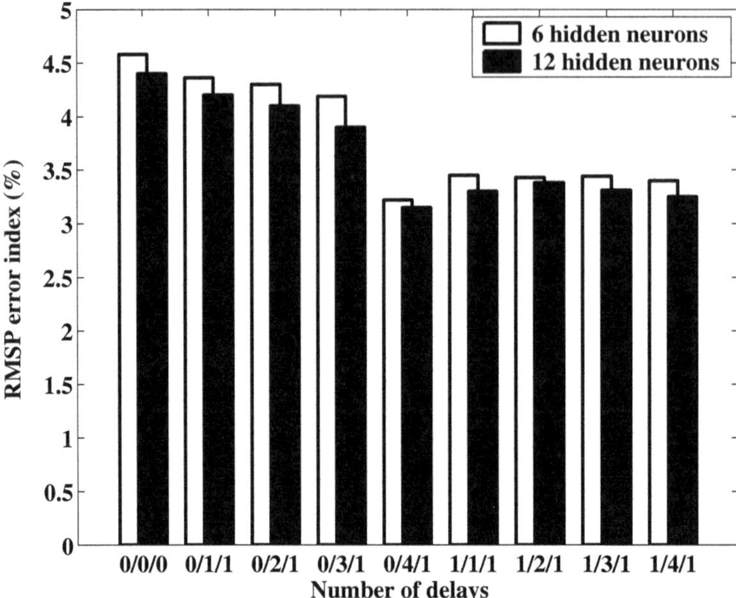

Fig. 4.4. Estimation root mean squared percentage error on testing data sets for neural networks with different combinations of delays. '0/4/1' indicates that input delay is zero, the number of output feedback delays are 1, 2, 3, and 4, the number of activation feedback delay is 1.

One of the simulation results of biomass prediction is plotted in Figure 4.5. The feed rate profile was saw-wave (see Figure 4.3(b)). The softsensor provided a good prediction of the growth of biomass with high fidelity. The prediction error showed oscillations occurring at the initial phase. This happened because the input delay was set to be zero. Previous inputs were not incorporated into the network, only the current inputs were presented for prediction. However, with the activation feedback and the output (estimated biomass concentration) feedback, the network just took a few iteration steps to settle down and then was able to move along the right track. From the prediction error in Figure 4.5, one can also see the prediction offset is small. The maximum percentage error of prediction is less than 3%.

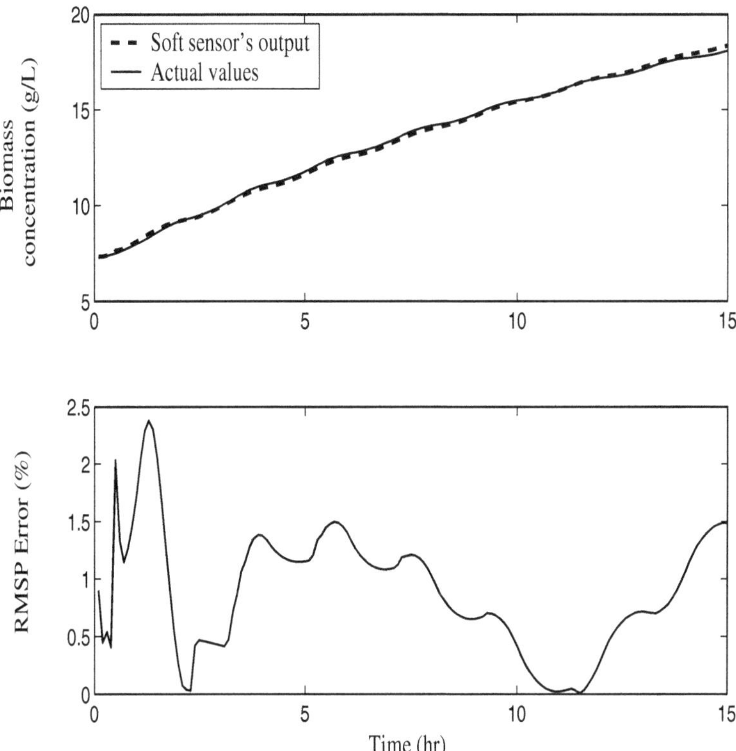

Fig. 4.5. Simulation result of softsensor using six hidden neuron network for a fed-batch fermentation process.

4.3 Experimental Verification

The main objective of these experiments is to investigate the workability of the proposed neural softsensor in real fermentation processes. In this study, three laboratory fed-batch cultures of baker's yeast have been carried out using a bench-scale fermentor.

Reactor setup

The schematic diagram of the experimental set-up is shown in Figure 4.6. Fed-batch cultivation was carried out in a BioFlo 3000 bench-top fermentor (New Brunswick Scientific Co., INC., USA). The vessel of the bioreactor has a total volume of 3L with a working volume of 2.5L. The input to the fermentor is the feed flow rate, which is controlled by a peristaltic pump. The measured outputs are the concentration of biomass and DO. The former is

measured by sampling the reaction broth; the latter is continuously monitored with an oxygen electrode (12mm, A-Type, Mettler-Toledo Process Analytical,Inc., USA). AFS-BioCommand process management software (New
Brunswick Scientific Co., INC., USA) is installed on a PC-compatible computer to set up the feeding trajectory prior to the fermentation. It is also used
to acquire data from sensors during the fermentation. A data communication
device (AFS-BioCommand Interface module, New Brunswick Scientific Co.,
INC., USA) is used to bridge between the computer and the bioreactors. The
temperature, agitation speed, airflow and pH are controlled through an on-
board proportional-integral-derivative (PID) controller, and they can also be
altered by using the software.

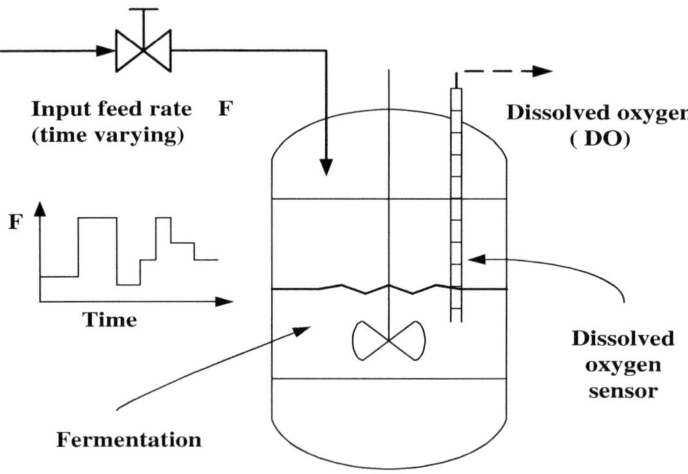

Fig. 4.6. Schematic diagram of fed-batch fermentation setup.

Experimental procedure

Yeast strain, *Saccharomyces cerevisiae*, produced by Goodman Fielder Milling
& Baking N.Z. Ltd. was grown in Yeast Extract, Peptone and Dextrose
(YEPD) medium [97] with the following composition: Dextrose, 20g/L; Yeast
extract, 10g/L; Peptone, 20g/L and commercial anti-foam, 10 drops/L. The
starter culture was created in a shaker at 30°C and 200RPM for 60 to 90
minutes.

The reactor vessel and initial medium were prepared prior to fed-batch
fermentation. 1.4 liter YEPD of liquid was added to the vessel as an initial
medium. The complete assembly of vessel, head-plate, pH probe and DO probe
were sterilized together with the initial medium by autoclave at 110°C for 20

minutes. The pH probe (Ingold Electrodes Inc.) was calibrated before the sterilization; while the DO probe (Mettler-Toledo Process Analytical, Inc.) was calibrated after the sterilization.

After the starter culture, 100 mL of pre-culture was inoculated from flasks into the vessel of bench-top fermentor. The temperature, agitation speed, airflow and pH were controlled at the nominal values of $30°C$, 800RPM, 3L/min and 4.5 respectively. Nutritive substrate was automatically added into the bioreactor according to the predetermined feed rate trajectory. Silicon tubes (HV-96400-16 Precision silicone (peroxide) Tubing, Masterflex, USA) with inside diameter 3.1mm were used to feed the nutrients to the reactor. The feed rate was pre-calibrated before the fermentation was started. Three different types of trajectories were used in the experiments: constant feed rate, square-wave feed rate and stair-shape feed rate. The total feed volume was 1L.

Biomass concentration was measured off-line, which involved measuring the wet weight of yeast after centrifuging (Eppendorf centrifuge, Germany) 10mL broth samples for 10 minutes at 4500rpm and decanting the supernatant liquid. DO in the bioreactor was monitored by the oxygen electrode and the data was stored in a database in the computer. The fermentation volume was obtained by solving Equation B.6 as shown in Appendix B, and no measurement was needed.

The fermentor was operated for five to eight hours. Medium samples were taken approximately every 18 minutes to determine biomass concentrations, while DO values were monitored every minute. Three sets of data were collected. For the cultivation with constant feed rate, 14 biomass samples were obtained during a five-hour fermentation run; For the cultivations with square-wave and stair-shape feed rates, 27 biomass samples were measured during an eight-hour fermentation run. The experimental data are shown from Figure 4.7 to Figure 4.9. These three data sets were used for re-training the neural network, validating the network being trained and testing the prediction ability of the trained network respectively. Due to the infrequent sampling of biomass and unequal sampling time between DO and biomass, an interpolation method was needed to process the biomass data. To preserve the monotonicity and the shape of the data, a piecewise cubic interpolation method [98] was adopted here.

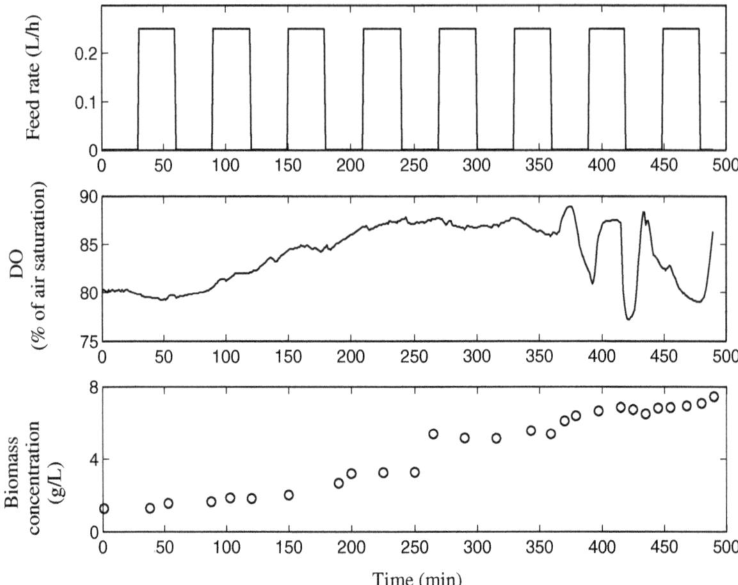

Fig. 4.7. Square-wave feed.

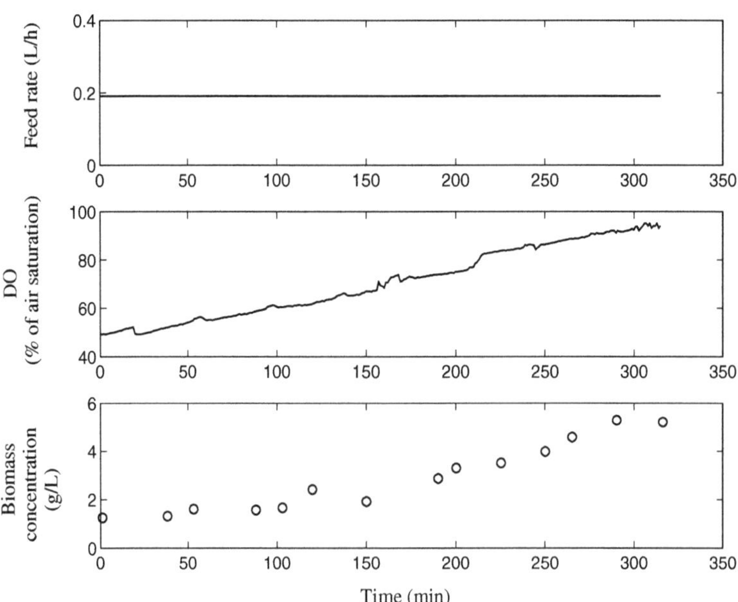

Fig. 4.8. Constant feed.

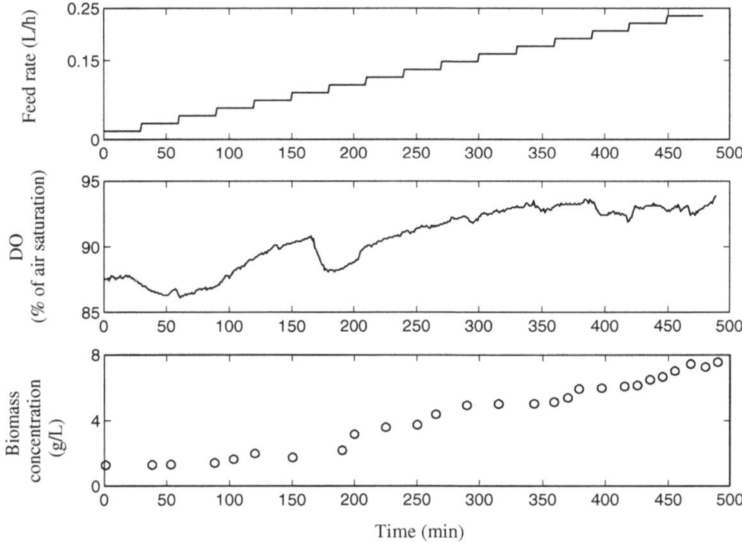

Fig. 4.9. Stair-shape feed.

Prediction results and discussion

Based on the simulation results obtained in Section 4.2, the network topology was optimized to the structure of 13-6-1 with zero input delay, one activation feedback delay and four output feedback delays. The 13 inputs to the input layer consisted of current states of feed rate, DO, volume, six activation feedbacks and four output feedbacks. This network structure was considered as a starting point for the experimental investigation.

Figure 4.10 shows the on-line biomass prediction when applying the neural softsensor to the unseen experimental data of stair-shape feed flow (see Figure 4.9). The prediction starts from an arbitrary initial point. As can be seen from the figure, the softsensor is able to converge within a very short time and can predict the trend of the growth of biomass. However, the MSE between the estimated values and the actual values of biomass is 0.3580. It is a little higher than 0.35, which has been previously reported in the literature by using Knowledge Based Modular networks [99]. As can be seen in the plot, a fluctuation appears in the prediction trajectory. The RMSP error is 11.7%. The network topology chosen for this prediction is exactly the same as the one used for the simulation given in Figure 4.5. As discussed in the simulation study, the main reason for the fluctuation could be that the historical input values are not presented to the network. Furthermore, under realistic conditions, errors in the biomass measurement, the effects of sampling, bias in the noise characteristics, noisy training data and batch to batch variations may have a significant affect on the estimation accuracy [73].

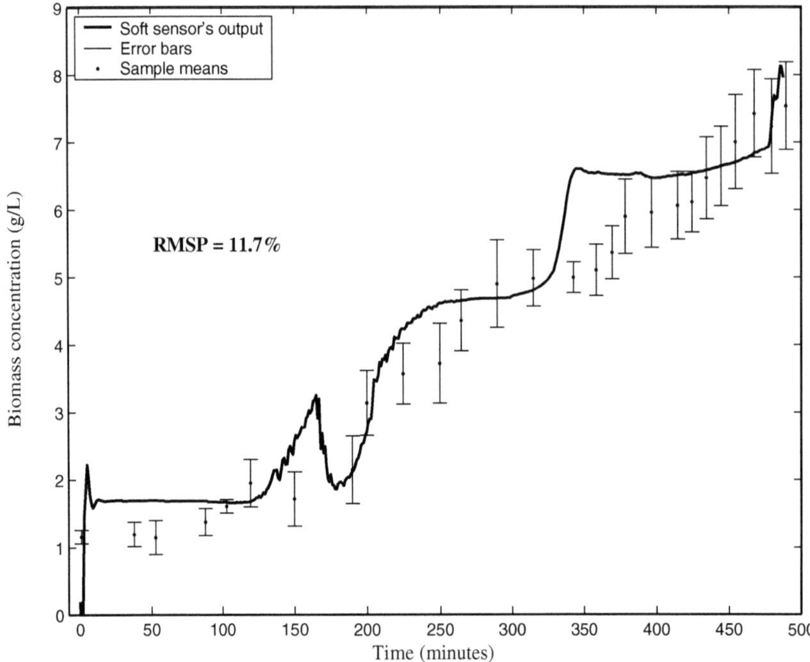

Fig. 4.10. On-line biomass concentration prediction in a fed-batch baker's yeast fermentation process. The network input delays: 0; output feedback delays: 1, 2, 3, 4; activation feedback delay: 1.

For the neural softsensor to overcome the output fluctuation, historic input values are required so that a smooth prediction, which is closer to the reality of biomass growth, can be achieved. Figure 4.11 shows the prediction result using a modified neural model in which two input delays have been incorporated through TDLs. In order to distinguish the effects caused by activation feedback delays and input delays, the activation feedback delays have been set to be zero. It is obvious that the fluctuation has been reduced significantly. However, the error is slightly higher than that in Figure 4.10 (RMSP error is 12.1%). In particular, the errors on the prediction at both ends, the beginning and the final period of the fermentation, are still large.

In order to improve the predictive ability on both ends of fermentation, an approach that was used in the study is to connect activation feedback to the network input through TDLs. After such modification, one can see from Figure 4.12 that a smooth prediction has been gained on both the initial phase and ending phase. The prediction RMSP error between the measured values and the estimated values is further decreased to 10.3%. Figure 4.10 to Figure 4.12 show a gradual improvement is achieved.

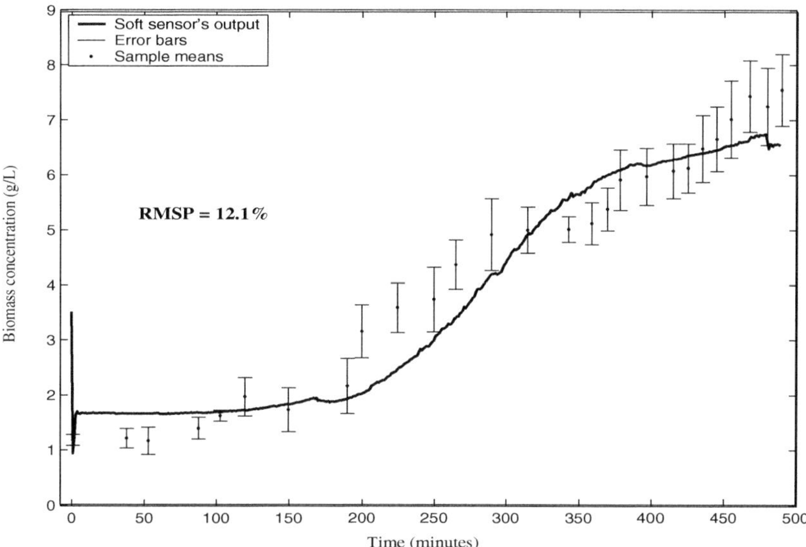

Fig. 4.11. On-line biomass concentration prediction in a fed-batch baker's yeast fermentation process. The network input delays: 1, 2; output feedback delays: 1, 2, 3, 4; activation feedback delay: 0.

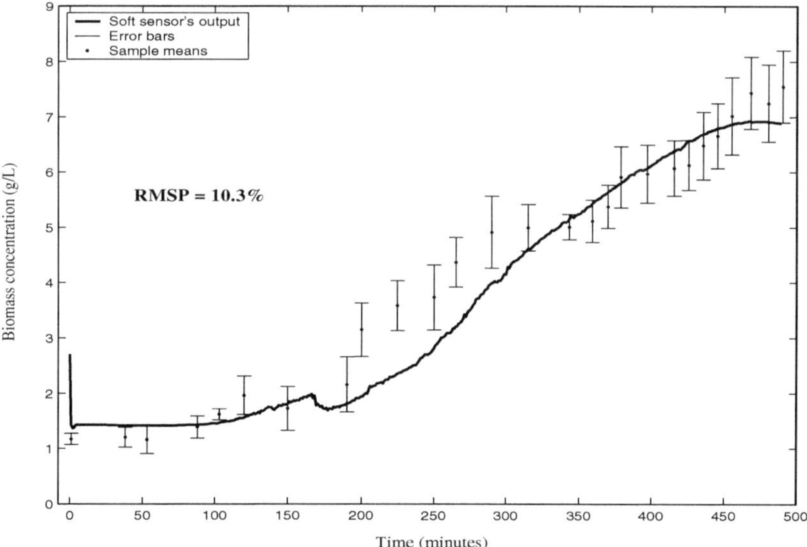

Fig. 4.12. On-line biomass concentration prediction in a fed-batch baker's yeast fermentation process. The network input delays: 1, 2; output feedback delays: 1, 2, 3, 4; activation feedback delay: 1.

For a comparison, the RMSP error and MSEs of the biomass predictions using three different topologies are listed in Table 4.1. The experimental results show that the highest predictive ability is obtained from the neural softsensor with two input delays, four output feedback delays and one activation feedback delay.

Table 4.1. Prediction errors using three different RNN topologies.

RNN topology	MSE	RMSP error(%)
0/1/4	0.3580	11.7
2/0/4	0.3159	12.1
2/1/4	0.3107	10.3

4.4 Conclusions

This work assesses the suitability of using RNNs for on-line biomass estimation in fed-batch fermentation processes. The proposed neural network sensor only requires the DO, feed rate and volume to be measured. Based on a simulated model, the neural network topology is selected. Simulations show that the neural network is able to predict the biomass concentrations within 3% of the true values. This prediction ability is further investigated by applying it to a laboratory fermentor. The experimental results show that the lowest RMSP error is 10.3%. From the results obtained in both simulation and real processes, it can be concluded that RNNs are powerful tools for on-line biomass estimation in fed-batch fermentation processes.

5

Optimization of Fed-batch Fermentation Processes using Genetic Algorithms based on Cascade Dynamic Neural Network Models

A combination of a cascade RNN model and a modified GA for optimizing a fed-batch bioreactor is investigated in this chapter. The complex nonlinear relationship between the manipulated feed rate and the biomass product is described by two recurrent neural sub-models. Based on the neural model, the modified GA is employed to determine a smooth optimal feed rate profile. The final biomass quantity yields from the optimal feed rate profile based on the neural network model reaches 99.8% of the "real" optimal value obtained based on a mechanistic model.

5.1 Introduction

Mechanistic models are conventionally used to develop optimal control strategies for bioprocesses [100, 101, 102, 103]. However, to obtain a mechanistic model for bioprocesses is a time-consuming and costly work. The major challenge is the complex and time-varying characteristics of such processes.

In Chapter 4, a softsensor is proposed using RNN for predicting biomass concentration from the measurement of DO, feed rate and volume. In this chapter, we intend to model the fed-batch fermentation of *Saccharomyces cerevisiae* from the input of feed rate to the output of biomass concentration by cascading two softsensors.

An example of a recurrent dynamic neural network is illustrated in Figure 1.4 in Chapter 1. In this structure, besides the output feedback, the activation feedbacks are also incorporated into the network, and TDLs are used to handle the delays. A dynamic model is built by cascading two such extended RNNs for predicting biomass concentration. The aim of building the neural model is to predict biomass concentration based purely on the information of the feed rate. The model can then be used to maximize the final quantity of biomass at the end of the reaction time by manipulating the feed rate profiles.

This chapter is organized as follows: in Section 5.2, the mechanistic model of industrial baker's yeast fed-batch bioreaction is given; in Section 5.3, the

L. Z. Chen et al.: *Modelling and Optimization of Biotechnological Processes*, Studies in Computational Intelligence (SCI) **15**, 57–70 (2006)
www.springerlink.com

development of the cascade RNN model is presented; Section 5.4 shows the results of neural model prediction; the optimization of feed rate profile using the modified GA is described in Section 5.5; Section 5.6 summarizes this chapter.

5.2 The Industry Baker's Yeast Fed-batch Bioreactor

The mathematical model, which consists of six differential equations [17, 92], was used to generate simulation data. The details of the model parameters and initial conditions are given in Appendix B. Three output variables, biomass, DO and volume were generated from a given feed rate by solving the six differential equations:

$$
\begin{aligned}
\frac{d(V \cdot C_s)}{dt} &= F \cdot S_0 - \left(\frac{\mu}{Y_{x/s}^{ox}} + \frac{Q_{e,pr}}{Y_{e/s}} + m\right) \cdot V \cdot X \\
\frac{d(V \cdot C_o)}{dt} &= -Q_o \cdot V \cdot X + k_L a_o \cdot (C_o^* - C_o) \cdot V \\
\frac{d(V \cdot C_c)}{dt} &= Q_c \cdot V \cdot X + k_L a_c \cdot (C_c^* - C_c) \cdot V \\
\frac{d(V \cdot C_e)}{dt} &= (Q_{e,pr} - Q_{e,ox}) \cdot V \cdot X \\
\frac{d(V \cdot X)}{dt} &= \mu \cdot V \cdot X \\
\frac{dV}{dt} &= F
\end{aligned}
\tag{5.1}
$$

where, C_s, C_o, C_c, C_e , X, and V are state variables which denote concentrations of glucose, dissolved oxygen, carbon dioxide, ethanol, and biomass, respectively; V is the liquid volume of the fermentation; F is the feed rate which is the input of the system; m is the glucose consumption rate for the maintenance energy; $Q_{e,pr}$, Q_o, Q_c and $Q_{e,ox}$ are ethanol production rate, oxygen consumption rate, carbon dioxide production rate and oxidative ethanol metabolism, correspondingly; $Y_{e/s}$ and $Y_{x/s}^{ox}$ are yield coefficients; $k_L a_o$ and $k_L a_c$ are volumetric mass transfer coefficients; S_0 is the concentration of feed.

Five different feed rate profiles, which are shown in Figure 4.3 as given in Chapter 4, were chosen to generate training and testing data: (1) the square-wave feed flow, (2) the saw-wave feed flow, (3) the stair-shape feed flow, (4) the industrial feeding policy, (5) the random-steps feed flow.

5.3 Development of Dynamic Neural Network Model

Cascade dynamic neural network model

A dynamic neural network model is proposed in this study using a cascade structure as shown in Figure 5.1. It contains two extended recurrent neural blocks which model the dynamics from inputs, F and V, to the key variable C_o and the fermentation output (product) X. The first block estimates the trend of C_o which provides important information to the second neural block. The second neural block acts exactly as a softsensor developed in the researcher's previous work [86], which is described in Chapter 4, except that instead of the measured value of DO, the estimated value of DO is used here as the input of the second neural block. The softsensor model requires DO data measured on-line, whereas the cascade dynamic model proposed in Figure 5.1 basically needs only the data of the feed rate to predict the biomass concentration. Although the volume is another input for the model, it can be simply calculated by using Equation B.6 as shown in Appendix B.

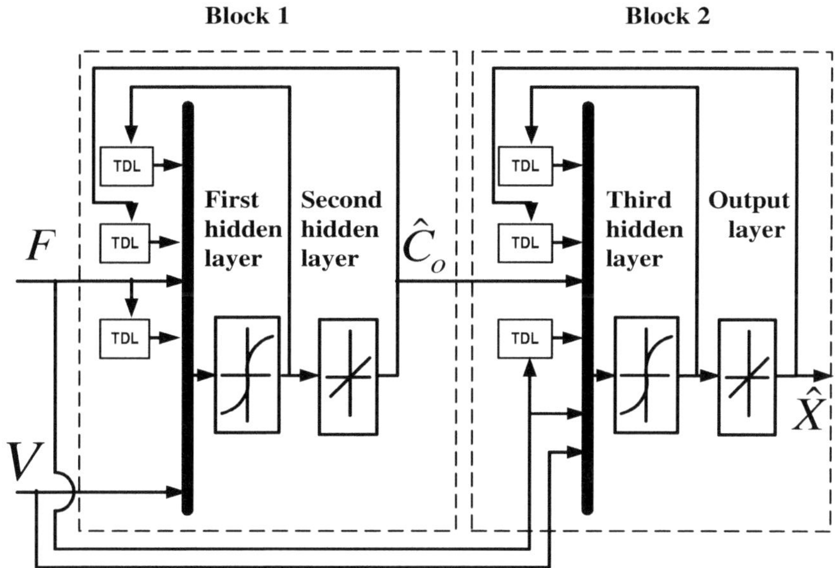

Fig. 5.1. Structure of the proposed recurrent neural model.

In each of the neural blocks, both feed-forward and feedback paths are connected through TDLs in order to enhance the dynamic behaviors. All connections could be multiple paths. Sigmoid activation functions are used for the hidden layers and a pure linear function is used for the output layers. The structure of the neural blocks reflects the differential relationships between inputs and outputs as given by Equation B.2 to Equation B.6. A full mathematical description of the cascade model is given in the following equations. The output of the i-th neuron in the first hidden layer is of the form:

$$h_{1i}(t) = f_1\left(\sum_{j=0}^{n_a} W_{ij}^I \, u_1(t-j) + \sum_{k=1}^{n_b} W_{ik}^R \, \hat{C}_o(t-k)\right.$$

$$\left. + \sum_{l=1}^{n_c} W_{il}^{H1} \, h_1(t-l) + b_i^{H1}\right) \tag{5.2}$$

where, u_1 and h_1 are the vector values of the neural network input and the first hidden layer's output, correspondingly; $\hat{C}_o$ is the second hidden layer output; b_i^{H1} is the bias of i-th neuron in first hidden layer; n_a, n_b, n_c are the number of input delays, the number of the second hidden layer feedback delays and the number of first hidden layer feedback delays, respectively; $f_1(\cdot)$ is a sigmoidal function; W_{ij}^{I} are the weights connecting the j-th delayed input to i-th neuron in the first hidden layer, W_{ik}^{R} are the weights connecting the k-th delayed second hidden layer output feedback to the i-th neuron in the first hidden layer, W_{il}^{H1} are the weights connecting the l-th delayed activation feedback to the i-th neuron in the first hidden layer.

Note that one neuron is placed at the output of the second hidden layer, so that:

$$\hat{C}_o(t) = f_2\Big(\sum_{m=1}^{n_g} W_m^{Y}\, h_m(t) + b_Y\Big) \tag{5.3}$$

where, $f_2(\cdot)$ is a pure linear function; W_m^{Y} are the weights connecting the m-th neuron in the first hidden layer to the second hidden layer; n_g is the number of neurons in the first hidden layer; b_Y is the bias of the second hidden layer.

The second neural block has an additional input, $\hat{C}_o$. Similar to the first block, the output of i-th neuron in the third hidden layer can be described as:

$$h_{3i}(t) = f_1\Big(\sum_{j=0}^{n_d} W_{ij}^{P}\, u_2(t-j) + \sum_{k=1}^{n_e} W_{ik}^{O}\, \hat{X}(t-k)$$
$$+ \sum_{l=1}^{n_f} W_{il}^{H3}\, h_3(t-l) + b_i^{H3}\Big) \tag{5.4}$$

where, u_2 and h_3 are the vector values of the input to the third hidden layer and the third hidden layer's output, correspondingly; $\hat{X}$ is the model's output; b_i^{H3} is the bias of i-th neuron in the third hidden layer; n_d, n_e, n_f are the number of input delays to the third hidden layer, the number of the output layer feedback delays and the number of third hidden layer feedback delays, respectively; $f_1(\cdot)$ is the sigmoidal function; W_{ij}^{P} are the weights connecting the j-th delayed input of the third hidden layer to the i-th hidden neuron in the layer, W_{ik}^{O} are the weights connecting the k-th delayed output feedback to the i-th neuron in the third hidden layer, W_{il}^{H3} are the weights connecting the l-th delayed activation feedback to the i-th neuron in the third hidden layer.

The model's output, which is the estimated biomass concentration can be expressed as:

$$\hat{X}(t) = f_2\Big(\sum_{m=1}^{n_k} W_m^{X}\, h_m(t) + b_X\Big) \tag{5.5}$$

where, $f_2(\cdot)$ is a pure linear function; W_m^{X} are the weights connecting the m-th neuron in the third hidden layer to the output layer; n_k is the number of neurons in the third hidden layer; b_X is the bias of the output layer.

Neural network training

A schematic illustration of the neural network model training is shown in Figure 5.2. The output of the bioprocess is used only for training the network. The model predicts the process output using the same input as the process after training. No additional measurements are needed during the prediction phase.

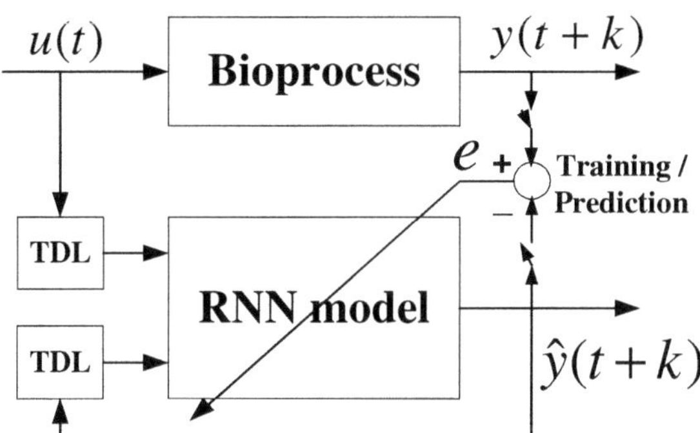

Fig. 5.2. Schematic illustration of neural network model training and prediction.

The goal of network training is to minimize the MSE between the measured value and the neural network's output by adjusting it's weights and biases. The LMBP training algorithm is adopted to train the neural networks due to its fast convergence and memory efficiency [34].

To prevent the neural network from being over-trained, an early stopping method is used here. A set of data which is different from the training data set (e.g., saw-wave) is used as a validation data set. The error on the validation data set is monitored during the training process. The validation error will normally decrease during the initial phase of training. However, when the network begins to over-fit the data, the error on the validation set typically begins to rise. When the validation error increases for a specified number of iterations, the training is stopped, and the weights and biases of the network at the minimum of the validation error are obtained.

The rest of the data sets, which are not seen by the neural network during the training period, are used in examining the trained network. The performance function that is used for testing the neural networks is the RMSP error index [57], which is defined in Equation 4.5.

A smaller error on the testing data set means the trained network has achieved better generalization. Two different training patterns, overall training and separated training, are studied. When the overall training is used, the whole network is trained together. When the separated training is used, block one and block two are trained separately. A number of networks with different numbers of hidden neuron delays are trained. For each network structure, 50 networks are trained; the one that produces the smallest RMSP error for the testing data sets is retained. The number of hidden neurons for the first hidden layer and the third hidden layer are 12 and 10 respectively. Errors for different training patterns and various combinations of input and feedback delays are shown in Figure 5.3. As shown in this figure, the 6/4/4 structure (the feed rate delays are six, the first block output delays and the second block output delays are four) has the smallest error and is chosen as the process model. The separated training method is more time-consuming but is not superior to the overall training. Thus, the overall training is chosen to train the network whenever new data is available.

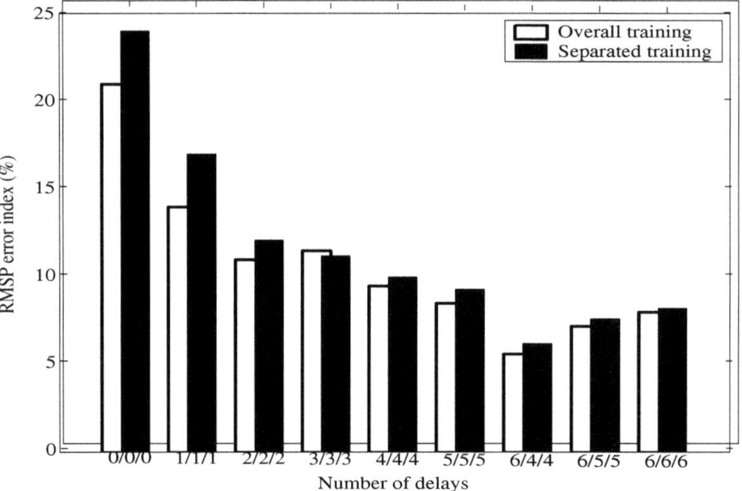

Fig. 5.3. Biomass prediction error on testing data sets for neural models with different combinations of delays. '6/4/4' indicates that the number of feed rate delays is six; the number of the first block output feedback delays is four; and the number of the second block output feedback delays is four.

5.4 Biomass Predictions using the Neural Model

The biomass concentrations predicted by the neural network model and the corresponding feed rates and prediction errors are plotted in Figures 5.4 to 5.6. As shown in these figures, the prediction error is quite big at the initial period of fermentation and gradually becomes smaller and smaller. The prediction error is less than 8%.

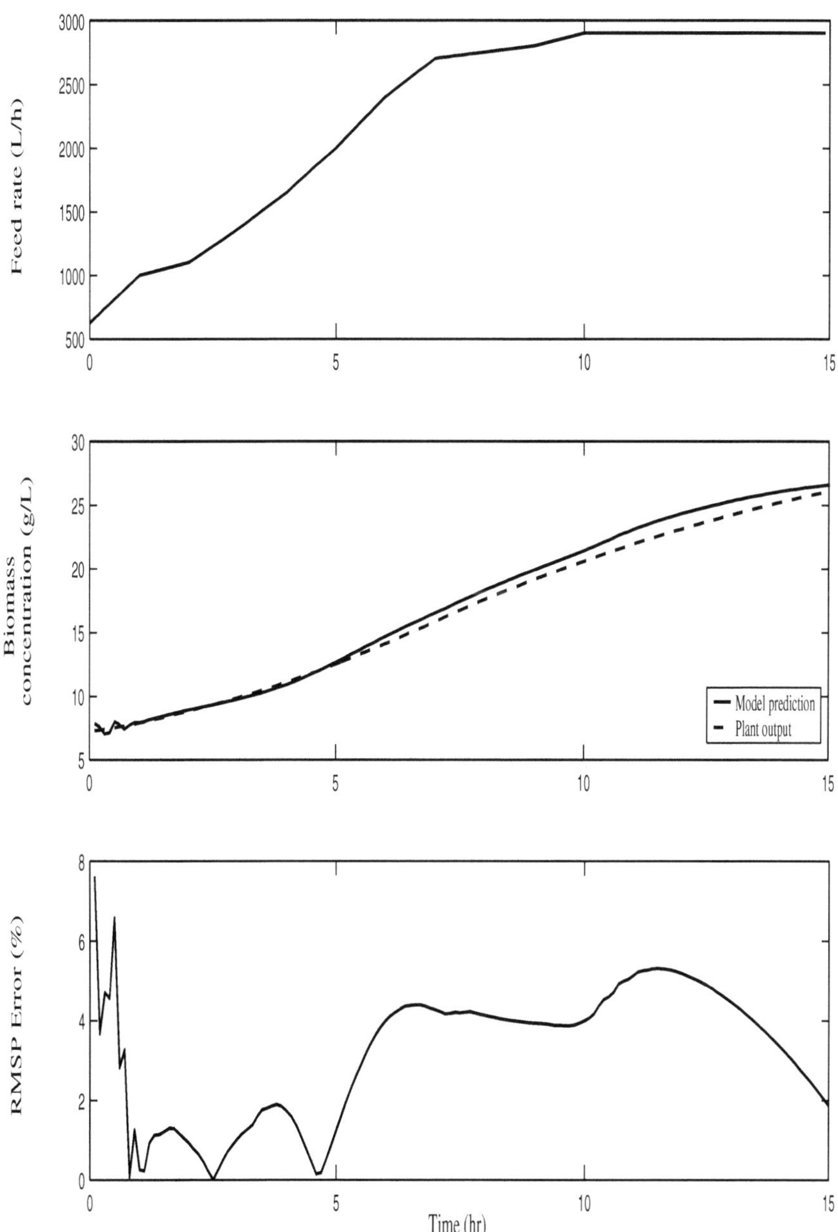

Fig. 5.4. Biomass prediction for the industrial feed rate profile.

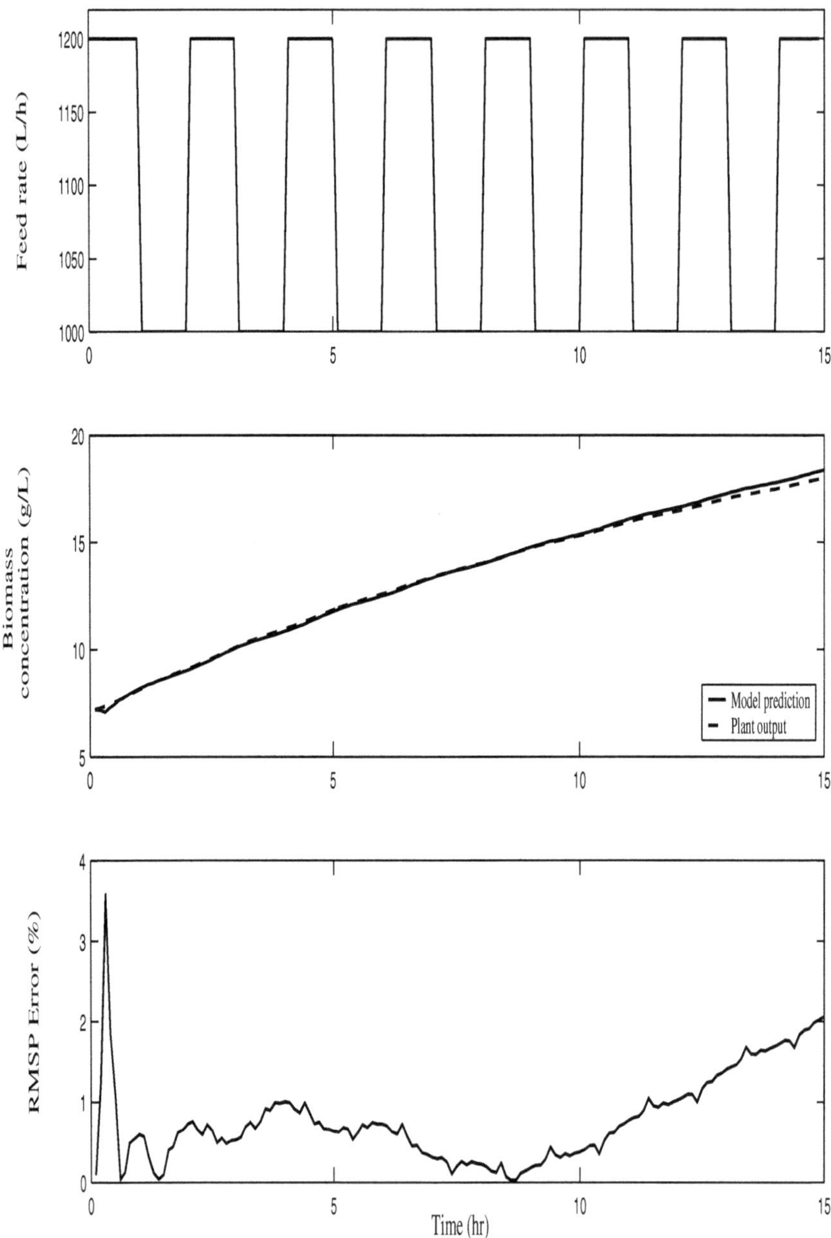

Fig. 5.5. Biomass prediction for the square-wave feed rate profile.

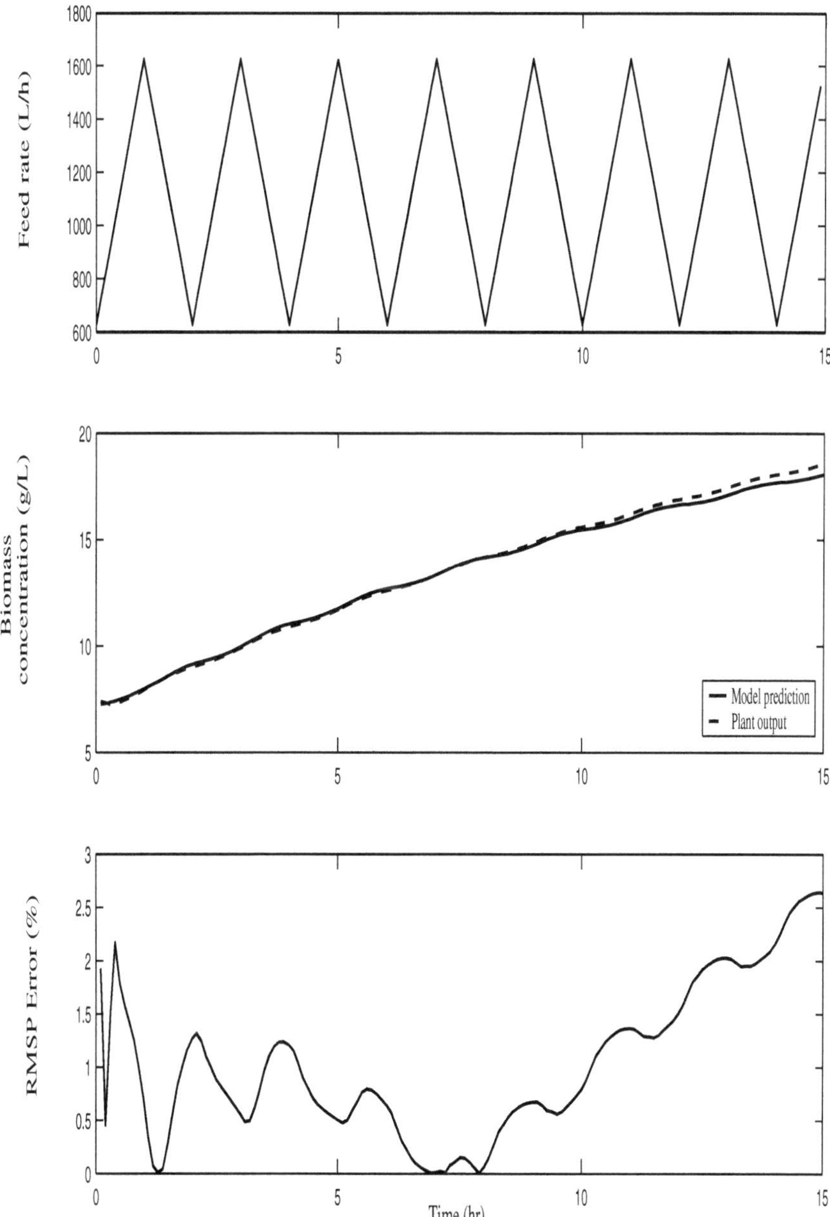

Fig. 5.6. Biomass prediction for the saw-wave feed rate profile.

5.5 Optimization of Feed Rate Profiles

Once the cascade recurrent neural model is built, it can be used to perform the task of feed rate profile optimization. The GA is used in this work to search for the best feed rate profiles.

GAs tend to seek for better and better approximations to a solution of a problem when running from generation to generation. The components and mechanism of GAs are described in Chapter 1 and 2. A simple standard procedure of a GA is summarized here by the following five steps: (i) Create an initial population of a set of random individuals. (ii) Evaluate the fitness of individuals using the objective function. (iii) Select individuals according to their fitness, then perform crossover and mutation operations. (iv) Generate a new population. (v) Repeat steps ii - iv until termination criteria is reached.

The feed flow rate, which is the input of the system described in Section 2, is equally discretized into 150 constant control actions. The total reaction time and the final volume are fixed to be 15 hours and 90,000 liters, respectively. The control vector of the feed rate sequence is:

$$F = [F_1 \quad F_2 \quad \cdots \quad F_{150}]^T \tag{5.6}$$

The optimization problem here is to maximize the amount of biomass quantity at the end of the reaction. Thus, the objective function can be formulated as follows:

$$\max_{F(t)} J = X(t_f) \times V(t_f) \tag{5.7}$$

where t_f is the final reaction time.

The optimization is subject to the constraints given below:

$$\begin{aligned} 0 \leq \quad F &\leq 3500 \ L/h \\ V(t_f) &\leq \quad 90000 \ L \end{aligned} \tag{5.8}$$

In this study, optimization based on the mathematical model is first performed to find the best feed rate profile and the highest possible final biomass productivity that can be obtained. Then the optimization is performed again using the RNN model. The resulting optimal feed rate is applied to the mathematical model to find the corresponding system responses and the final biomass quantity. As mentioned above, the mathematical model is considered here as the actual "plant". Thus, the suitability of the proposed neural network model can be examined by comparing these two simulation results.

The optimal profile that is obtained by using a standard GAs is highly fluctuating. This makes the optimal feed rate profile less attractive for practical use, because extra control costs are needed and unexpected disturbances may be added into the bioprocesses. In order to eliminate the strong variations on the optimal trajectory, the standard GA is modified. Instead of introducing new filter operators into the GA [80], a simple compensation method is integrated into the evaluation function. The control sequence F is amended inside the evaluation function to produce a smoother curve of feed trajectory while the evolutionary property of the GA is still maintained. This operation has no effect on the final volume.

The method includes three steps:

1. Calculate the distance between two neighboring individuals F_i and F_{i+1} using $d = |F_i - F_{i+1}|$, where $i \in (1, 2, \cdots, 150)$.
2. If d is greater than a predefined value (e.g., 10 L/h) then move F_i and F_{i+1} by $d/3$ towards the middle of F_i and F_{i+1} to make them closer.
3. Evaluate the performance index J for the new control variables.
4. Repeat steps 1-3 until all individuals in the population have been checked.

The Matlab GAOT software is used to solve the problem. The population size was chosen at 150. The development of the optimal feed rate profiles based on the mechanistic model and neural network model from the initial trajectory to the final shape is illustrated in Figure 5.7 and Figure 5.8. As the number of the generation increases, the feeding trajectory gradually becomes smoother and smoother, and the performance index, J, is also increased. The smoothing procedure works in a more efficient way for the mathematical model; it takes 2000 generations to obtain a smooth profile, while 2500 generations are needed to smooth the profile for the neural network model. This is due to the disturbance rejection nature of the RNN. A small alteration in feed rate is treated as a perturbation, thus the network is rather unsensitive to it.

The optimization results using the modified GA are plotted in Figure 5.9. The results based on the mass balance equations (MBEs) are shown from (a) to (e). As a comparison, the results based on the cascade RNN model are shown from (f) to (j). The responses of the bioreactor to the optimal feed rate based on the neural model are also calculated using the mechanistic model. It can be seen that the two optimal trajectories are quite different. However, the final biomass quantities yielded from the optimal profile based on the neural model is $281,956$ C-mol. This is 99.8% of the yield from the optimal profile based on the mathematical model. Furthermore, the reactions of glucose, ethanol and DO are very similar for both optimal profiles. As shown in the Figure, ethanol is first slowly formed and increased in order to keep the biomass production rate at a high value. In the ending stage of the fermentation, the residual glucose concentration is reduced to zero, and ethanol is consumed in order to make the overall substrate conversion into biomass close to 100%.

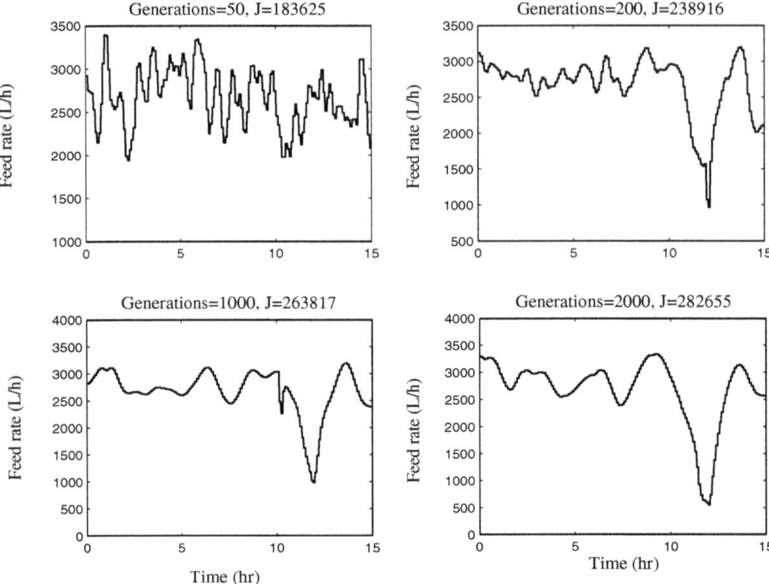

Fig. 5.7. Evolution of feed rate profile using the modified GA based on the mathematical model.

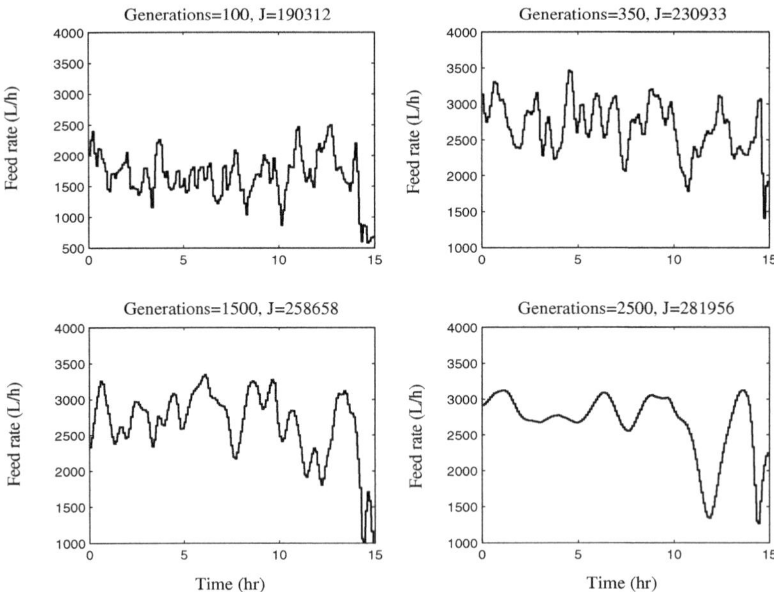

Fig. 5.8. Evolution of feed rate profile using the modified GA based on the RNN model.

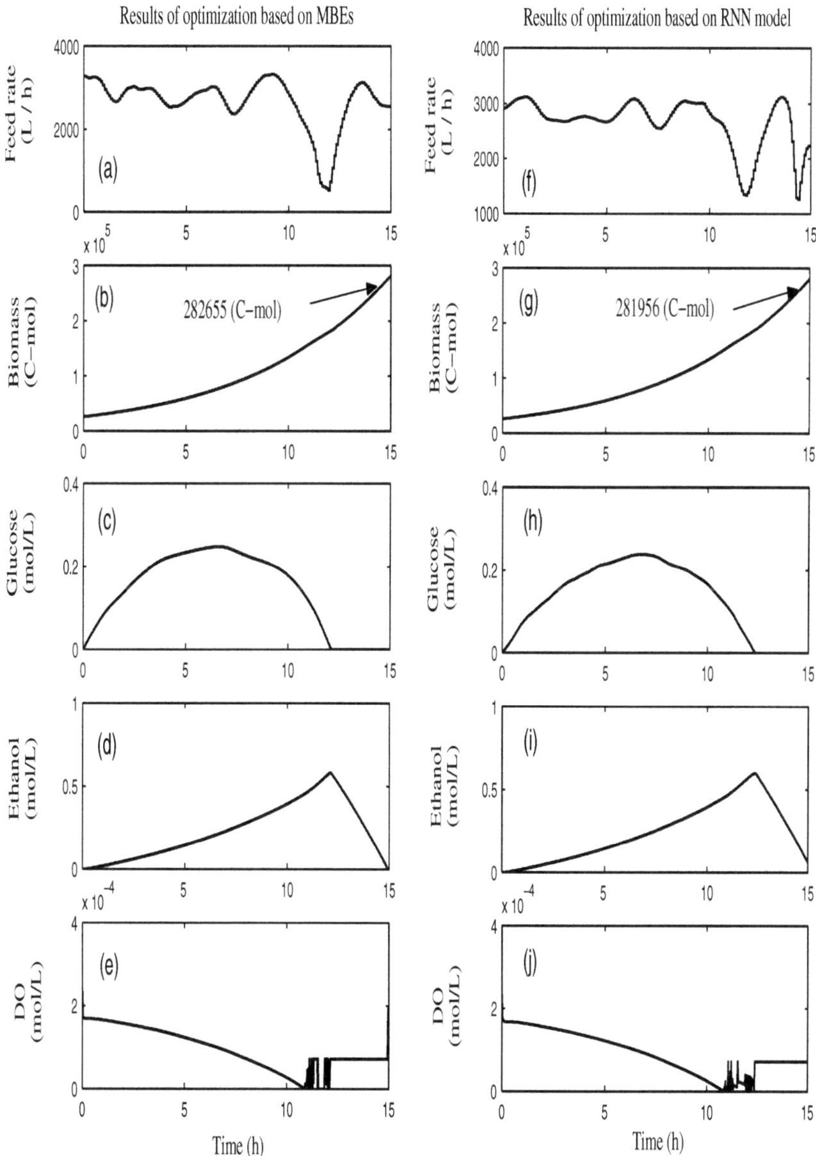

Fig. 5.9. Comparison of optimization results based on the mathematical model and RNN model.

5.6 Summary

Optimization of a fed-batch bioreactor using the cascade RNN model and the modified GA is investigated in this simulation study. The complex nonlinear relationship between manipulated feed rate and biomass product is represented by two neural blocks, in which outputs of one block are supplied into another neural block to provide meaningful information for biomass prediction. The results show that the error of prediction is less than 8%. The proposed model proves capable of capturing the dynamic nonlinear underlying phenomena contained in the training data set. The feasibility of the neural network model is further tested through the optimization procedure using the modified GA, which provides a mechanism to smooth feed rate profiles. The comparison of results of optimal feeding trajectories obtained based both on the mechanistic model and the neural network model have demonstrated that the cascade recurrent neural model is competent in finding the optimal feed rate profiles. The evolution of feed rate profiles through generations shows that the modified GA is able to generate a smooth profile, while the optimality of the feed rates is still maintained. The final biomass quantity yields from the optimal feeding profile based on the neural network model reaches 99.8% of the "real" optimal value.

6

Experimental Validation of Cascade Recurrent Neural Network Models

This chapter examines cascade RNN models for modelling bench-scale fed-batch fermentation of *Saccharomyces cerevisiae*. The models are experimentally identified through training and validating using the data collected from experiments with different feed rate profiles. Data preprocessing methods are used to improve the robustness of the neural network models. The results show that the best biomass prediction ability is given by a DO cascade neural model.

6.1 Introduction

A large number of simulation studies of neural network modelling have been reported in the literature [104, 105, 106, 107], and good results have generally been obtained. However, only a few of such studies have been taken the further step to experimental validation. Simulations allow systematic study of the complex bioreaction without conducting real experiments. However, because of the inherent nonlinear dynamic characteristics of fermentation processes, the process-model mismatching problem could significantly affect the accuracy of the results.

The main objective of this study is to model a laboratory scale fed-batch fermentation by neural network models using the cascade recurrent structure proposed in Chapter 5.

The remaining sections of this chapter proceed as follows: in Section 6.2, the cascade RNN and mathematical models are given; in Section 6.3, the experimental procedure is described; in Section 6.4, the experimental model identification and various aspects of data processing are detailed; conclusions are drawn in Section 6.5.

L. Z. Chen et al.: *Modelling and Optimization of Biotechnological Processes*, Studies in Computational Intelligence (SCI) **15**, 71–89 (2006)
www.springerlink.com

6.2 Dynamic Models

Recurrent neural network models

Two neural network models employed in this work are shown in Figure 6.1 and Figure 6.2. Development of these kinds of neural models is described in Chapter 5. The difference between Figure 6.1 and Figure 6.2 is that model I uses C_o, which is the concentration of DO, as its state variable, while model II uses the concentration of glucose C_s as its state variable.

Both model I and II use cascade structures, which contain two recurrent neural blocks. They model the dynamics from inputs, F and V, to the key variable C_o (or C_s) and the biomass concentration X. The first block estimates the trend of C_o (or C_s) which provides important information to the second neural block. The topology of each neural block is the same as that of the softsensor developed in Chapter 4.

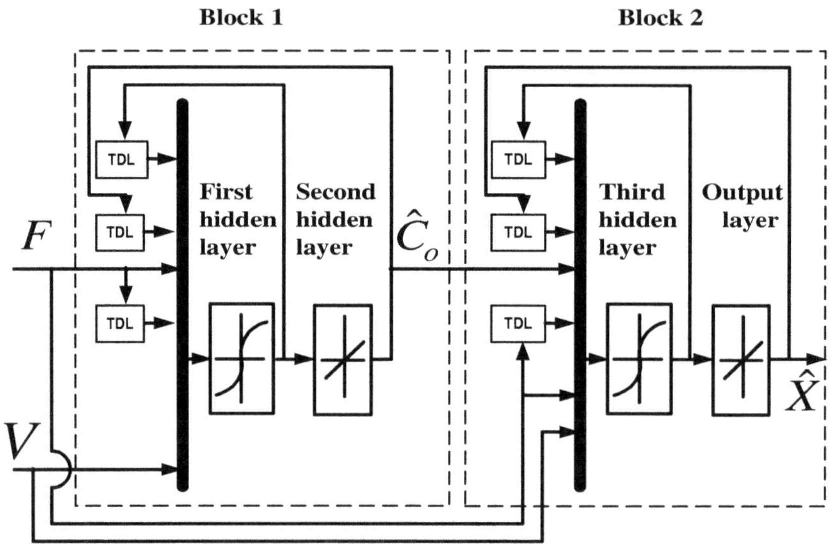

Fig. 6.1. Cascade RNN model I: DO RNN model.

In each of the neural blocks, both feed-forward and feedback paths are connected through TDLs in order to enhance the dynamic behaviors. Sigmoid activation functions are used for the hidden layers and a pure linear function is used for the output layers. The structure of the neural blocks reflects the differential relationships between inputs and outputs.

The first neural block can be described as follows:

$$\hat{C}(t+1) = f_1(\hat{C}(t), \hat{C}(t-1), \cdots, \hat{C}(t-m), H_1(t), H_1(t-1), \cdots, H_1(t-u),$$
$$F(t), F(t-1), \cdots, F(t-n), V(t)) \tag{6.1}$$

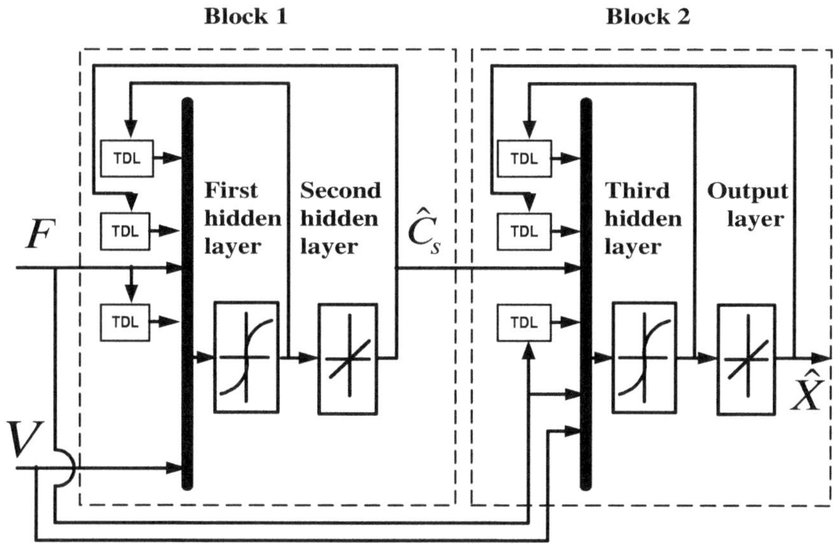

Fig. 6.2. Cascade RNN model II: Glucose RNN model.

where, $f_1(\cdot)$ is the nonlinear function represented by the first block; H_1 is a vector of the values of activation feedback in block 1; $\hat{C}$ represents $\hat{C}_o$ or $\hat{C}_s$, which is the concentration of DO or glucose; u, m and n are the maximum number of activation feedback delays, output layer feedback delays and input F delays in the first block correspondingly.

The second neural block has an additional input, $\hat{C}_o$ in Figure 6.1 or $\hat{C}_s$ in Figure 6.2, as compared with the first block. The predicted biomass concentration can be described as:

$$\hat{X}(t+1) = f_2(\hat{X}(t), \hat{X}(t-1), \cdots, \hat{X}(t-p), \hat{C}(t), H_2(t), H_2(t-1), \cdots,$$
$$H_2(t-v), F(t), F(t-1), \cdots, F(t-n), V(t)) \tag{6.2}$$

where, $f_2(\cdot)$ is the nonlinear function represented by the second block; H_2 is a vector of the values of activation feedback in block 2; v, p and n are the maximum number of activation feedback delays, output layer feedback delays and input F delays in the second block correspondingly. In this study, m, n, p, u, v are chosen as 6, 4, 4, 1, 1 respectively.

Dynamic mathematical model

To have a comparison with the neural models, a mathematical model was also identified for optimization. In this experimental investigation, four state variables, the concentration of biomass, DO, glucose and the fermentation volume were available. For a mathematical model that can describe the fermentation system with those available information, a popular mass balance equation structure, using simple Monod-like kinetics [108], was chosen. The required number of independent state variables was exactly the same as model II described above. The mass balance equations are in the form of:

$$\frac{dX}{dt} = \mu(S) - \frac{F}{V}X \tag{6.3}$$

$$\frac{dS}{dt} = -\frac{1}{Y_{XS}}\mu(S)X - mX + \frac{F}{V}(S_F - S) \tag{6.4}$$

$$\frac{V}{dt} = F \tag{6.5}$$

where, $\mu(S) = \frac{\mu_{max}S}{K_s + S + S^2/K_I}$; X and S are respectively the concentrations of biomass and glucose; S_F is the glucose concentration in the feeding solution; V is the liquid volume in the fermentor and F is the volumetric feed rate; K_S, K_I, μ_{max}, Y_{XS} and m are the model parameters.

The following initial culture conditions and feed concentrations have been used:

$$\begin{aligned}
S(0) &= 0 \ \ g/L \\
V(0) &= 1.0 \ \ L \\
X(0) &= 2.4 \ \ g/L \\
S_F &= 56.56 \ \ g/L
\end{aligned} \tag{6.6}$$

The GA was used to identify the parameters. The details of the identification method are described in Chapter 3. The identified parameters are listed in Table 6.1.

Table 6.1. The identified parameters that are used in the work.

Parameter	Value
K_S	0.01811 g/L
K_I	40.42 g/L
μ_{max}	0.3928 h^{-1}
Y_{XS}	0.2086
m	0.0000128 g g^{-1} h^{-1}

6.3 Experimental Procedure

Yeast strain and preservation

A pure culture of the prototrophic baker's yeast strain, *Saccharomyces cerevisiae* CM52(MATα his3-△200 ura3-52 leu2-△1 lys2-△202 trp1-△63), was obtained from the Department of Biologic Science, The University of Auckland (Auckland, New Zealand). A serial transfer method [109] was used for preserving and maintaining yeast strains. The pure culture was sub-cultured on YEPD agar slopes (Bacteriological peptone: 20g/l; yeast extract: 10g/l; glucose: 20g/l; agar: 15g/l), which were autoclaved for 15 minutes at 121°C . These cultures were kept in an incubator at 30°C for 3 days. Then the transferred cultures were stored at 0-4°C . The pure culture can be kept in a fridge for 30 days, after that a re-transferring is normally required. The stock cultures that were used for the inoculum preparation are shown in Figure 6.3.

Fig. 6.3. Yeast strain obtained from the Department of Biologic Science, The University of Auckland, are sub-cultured on YEPD agar slopes.

Growth of inoculum

The preserved culture was initially revived by growth in YEPD medium with the following composition: peptone, 20g/l; yeast extract, 10g/l; glucose: 20g/l. A 250mL shaker-flask contained 100mL of YEPD medium. Both flask and medium were sterilized at 110°C for 20 minutes. Yeast cells on the surface of a refrigerated agar slope were washed into 100ml of sterilized YEPD medium and propagated on the digital incubator shaker (Innova 4000, New Brunswick Scientific Co.,Inc., USA), as shown in Figure 6.4, at 30°C and 250 rpm for 12 hours. 100ml of such culture was used as the inoculum for each fermentation experiment.

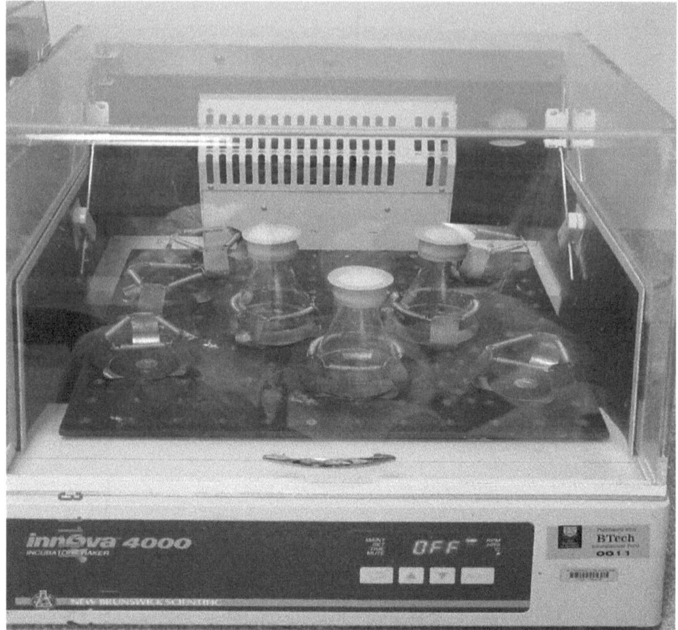

Fig. 6.4. Digital incubator shaker.

Batch and fed-batch phases

To allow the initial yeast inoculum to adapt to the new environment of bench-scale reactor and be sensitive to the feeding medium, a 12-hour batch fermentation was initially conducted after 100ml of inoculum was added into the reactor containing one liter YEPD medium. The medium was sterilized together with the reactor at 121°C for 25 minutes before a batch phase. During the batch phase, the temperature, agitation speed and air supply for the fermentation courses were respectively maintained at 30°C , 500 rpm and 4.0 L/min. The laboratory fermentor is shown in Figure 6.5.

The fed-batch fermentation was carried out under the same aerobic and temperature conditions of batch cultivation except a feeding medium was added in the bioreaction vessel, as shown in Figure 6.6, during the fed-batch cultivation. The feeding medium contained per liter: peptone, 200g; yeast extract, 100g; glucose, 200g; anti-foam, 20 drops. Due to the high glucose concentration in the feeding medium, the autoclave condition was changed to 110°C and 20 minutes. After sterilization, the glucose concentration was measured as 56.56 g/L. 1.5 liters of such medium was fed into the fermentor using the controllable peristaltic pump (illustrated in Figure 6.7) with flow rates between 0 and 0.2988 L/h.

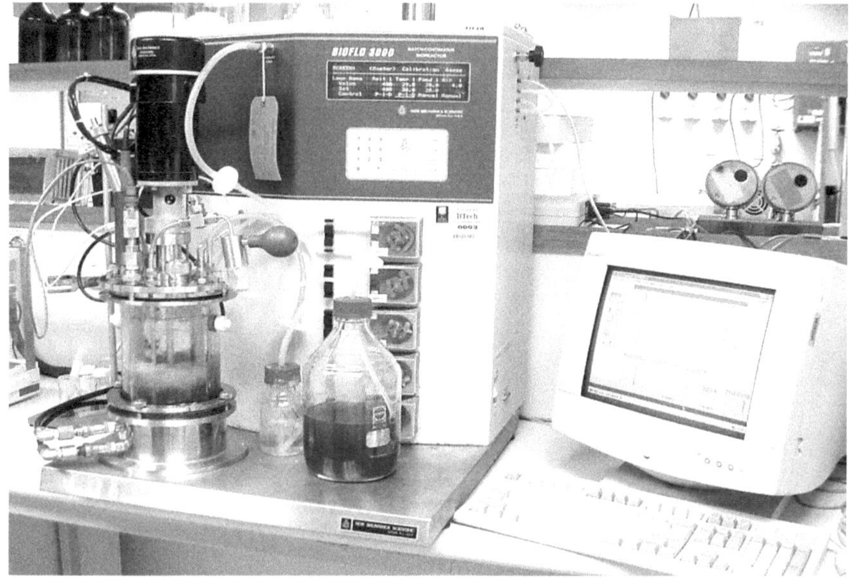

Fig. 6.5. Bench-scale fermentation equipment set-up. Model No.: BioFlo 3000 bench-top fermentor. Made by New Brunswick Scientific Co., INC., USA.

Fig. 6.6. Reaction vessel.

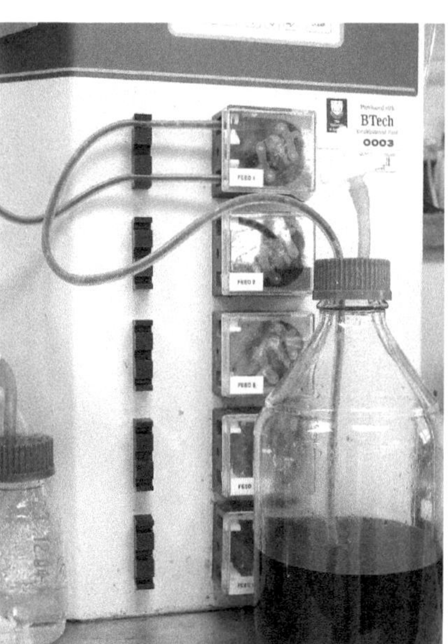

Fig. 6.7. Feeding control.

Sampling time

Fermentation processes were run for 12.5 hours for model identification and 8 hours for optimal feed rate validation. Medium samples were taken every 30 minutes approximately to determine biomass and glucose concentrations. The choice of sampling time was based on the practical operation condition and that suggested in the literature [110, 111]. The choice was guided by, but not restricted to, the dominant time constant of the process. A good rule of thumb is to choose the sampling time $\triangle t < 0.1\tau_{max}$, where τ_{max} is the dominant time constant. A 0.5-hour sampling time was chosen in this study. A total of 26 medium samples were measured during a 12.5-hour fermentation run.

Fig. 6.8. Sampling.

Analysis

Determination of culture dry weight

To compromise between the effects on the total volume of the bioreaction and the accuracy of the biomass concentration measurements, two 2-ml culture samples were taken from the broth (as shown in Figure 6.8) and centrifuged at 45000 rpm for five minutes (Eppendorf centrifuge, Germany) After decanting the supernatant liquid, distilled water was added to the tubes and it was centrifuged again to wash the residual medium components off the yeast cells. The wet yeast was dried at 100°C for 24 hours before being weighed using the balance (Sartorius BP110S, Sartorius AG GÖTTINGEN, Germany). The average dry weight was used for calculating the biomass concentration.

Measurement of glucose concentration

Glucose was measured off-line by reacting the glucose in glucose (Trinder) reagent (Sigma Diagnostics, USA) to yield a colored (red) solution. The change in color was measured by a spectrophotometer (Hewlett-Packard 8452A Diode Array Spectrophotometer, Germany) using a 505nm wavelength. The concentration was calculated by comparing the change in absorbency to a known standard glucose solution with the standard curve fitting error less than 2%.

Monitoring of dissolved oxygen

The values of DO were monitored on-line by the DO electrode as illustrated in Figure 6.9. Prior to measuring, the electrode was calibrated by frequently saturating it in the fermentation medium (without yeast) with air and by equating the instrument response with "air saturation". Instrument readings during the fermentation can then be expressed as "% of air saturation". The DO data were automatically logged into the computer every minute and stored in a process database.

Fig. 6.9. DO sensor.

6.4 Model Identification

Nine different feed rates for experimental identification

Due to the intensive data-driven nature of neural network modelling, a sufficient number of data are required. A data base that can provide "rich" enough information to build an appropriate and accurate input-output model is important [110].

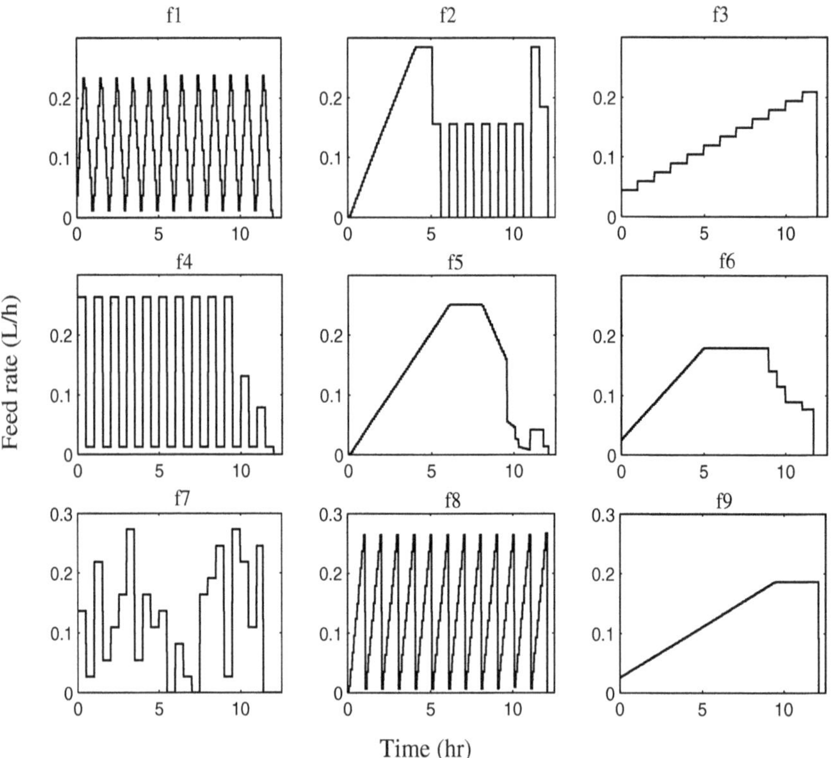

Fig. 6.10. Different feed rates for system identification.

In this study, a small data base was built by conducting nine experiments controlled by different feed rate profiles. The nine feed rates, which are illustrated in Figure 6.10, were designated in the experiments to excite the fed-batch fermentation system. They were carefully chosen in order to cover the "experimental space" as many times as possible and to yield informative data sets. Data which were collected during the bioreaction course were used to explore the complex dynamic behavior of the fed-batch fermentation system. They were used to train the neural networks and to identify the parameters of the mathematical model. The feed flow rate, measured data of DO, biomass, glucose and the calculated value of volume (using Equation 6.5) for one of the experiments are plotted in Figure 6.11.

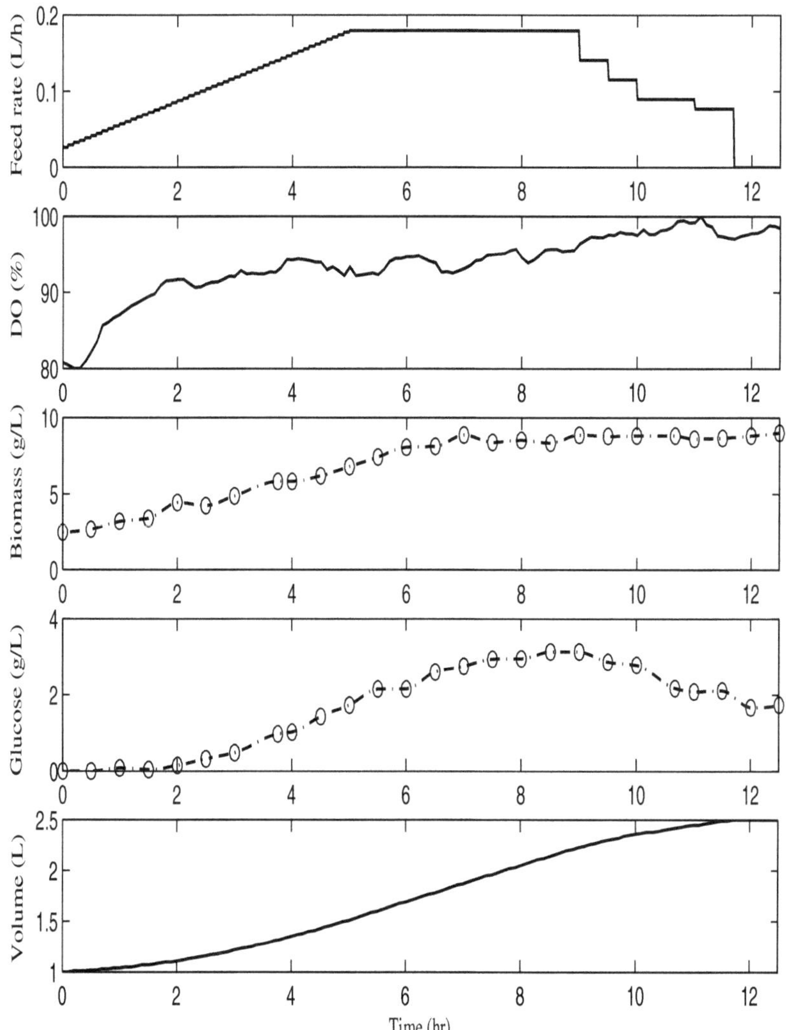

Fig. 6.11. One of the experimental data sets.

Training of neural network models

The aim of network training is to minimize the MSE between the measured value and the neural network's output. The LMBP training algorithm was employed to train the neural networks [34]. The explanation of neural network training and the LMBP algorithm are given in Chapter 4. In this section, a cross validation technique is emphasized.

An early stopping method is employed to prevent the neural network from being over-trained. A set of data which is independent from the training data sets is used as validation data. The error of validation is monitored during the training process. It will normally decrease during the initial phase of training. However, when the network begins to over-fit the data, the error on the validation set typically begins to rise. When the validation error increases for a specified number of iterations, the training is stopped, and the weights and biases of the network at the minimum of the validation error are obtained. The rest of the data, which is not seen by the neural network during the training and validation period, is used in examining the merit of the trained network.

Figure 6.12 shows the network training procedure. In this work, a total of nine experimental data sets were available. Seven of them were used for training, one of them was used for validation and one of them was used for testing. For each training, 50 networks were trained, and the one that generated the minimum test error was saved. Different combinations of the nine sets data were chosen in turn to train the network. The network that produced the minimum test error for all training was selected as the model of the fed-batch fermentation process. The number of hidden neurons for the first hidden layer and the third hidden layer were chosen as 12 and 10 respectively. The 6/4/4 structure (the feed rate delays are six, the first block output delays are four and the second block output delays are four) was selected as the topology of the network [112].

Data processing

Data interpolation

Due to the infrequent sampling of biomass and unequal sampling time between DO and biomass, an interpolation method was needed to process the experimental data before they could be applied to the model. To preserve the monotonicity and the shape of the data, a piecewise cubic interpolation method [98] was adopted in this study. After interpolation, the time step was 6 minutes for all data sequences .

DO data normalization

DO value was measured on-line and was recorded every one minute. From the data obtained, DO values were shown to be located between 20% and 100%.

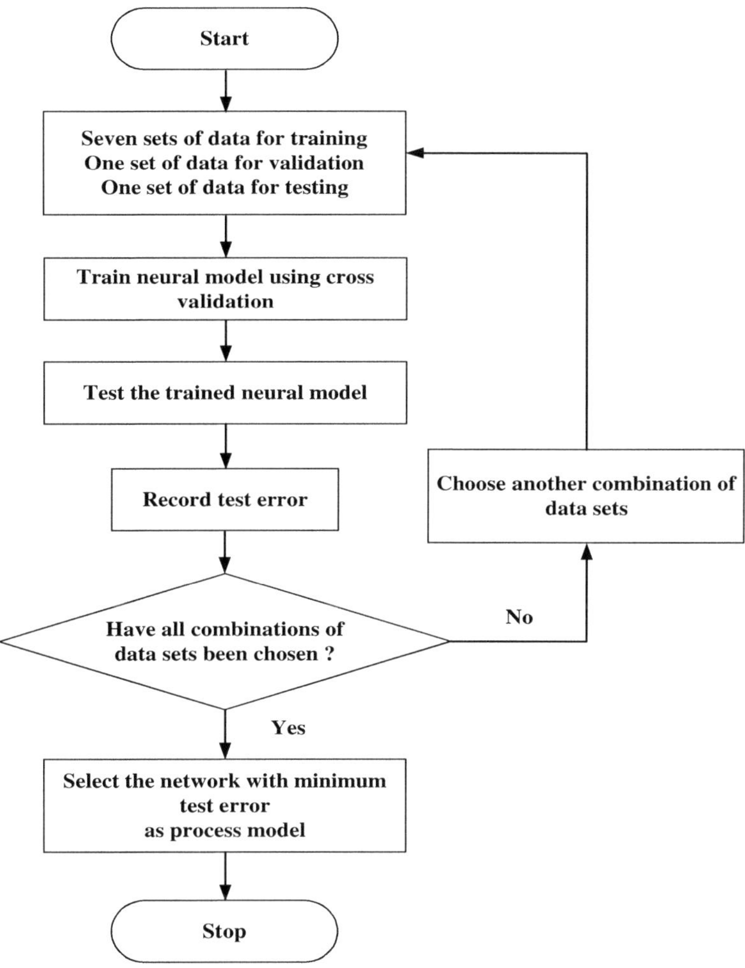

Fig. 6.12. Flowchart of the neural network training strategy.

But a few of the data sets were out of this range. This is due to the difficulties in calibration of the DO sensor, initial air saturations varying from batch to batch and the bubble formation during the bioreaction. The unexpected changes of DO range could significantly affect the accuracy of the biomasss concentration prediction, because the DO value is the key information for biomass estimation in the proposed neural network model I. Assuming that the trend of measured DO data was correct, a normalization method was used to bring the DO values that were outside the boundary to the range between 20% and 100%. The trend of the DO value was thus emphasized and the uncertainty on the biomass prediction was reduced. The mathematical formulation that was used for normalization is as follows:

$$C_{onew} = 80 \cdot (C_o - C_{omin})/(C_{omax} - C_{omin}) + 20 \qquad (6.7)$$

where, C_{onew}, C_o, C_{omin}, C_{omax} are normalized value, original value, maximum value and minimum value of original DO data, respectively.

Figure 6.13 shows the neural network predictions using the original data and the normalized data. It can be seen that a closer representation of the cell growth has been achieved by the model trained with normalized DO data.

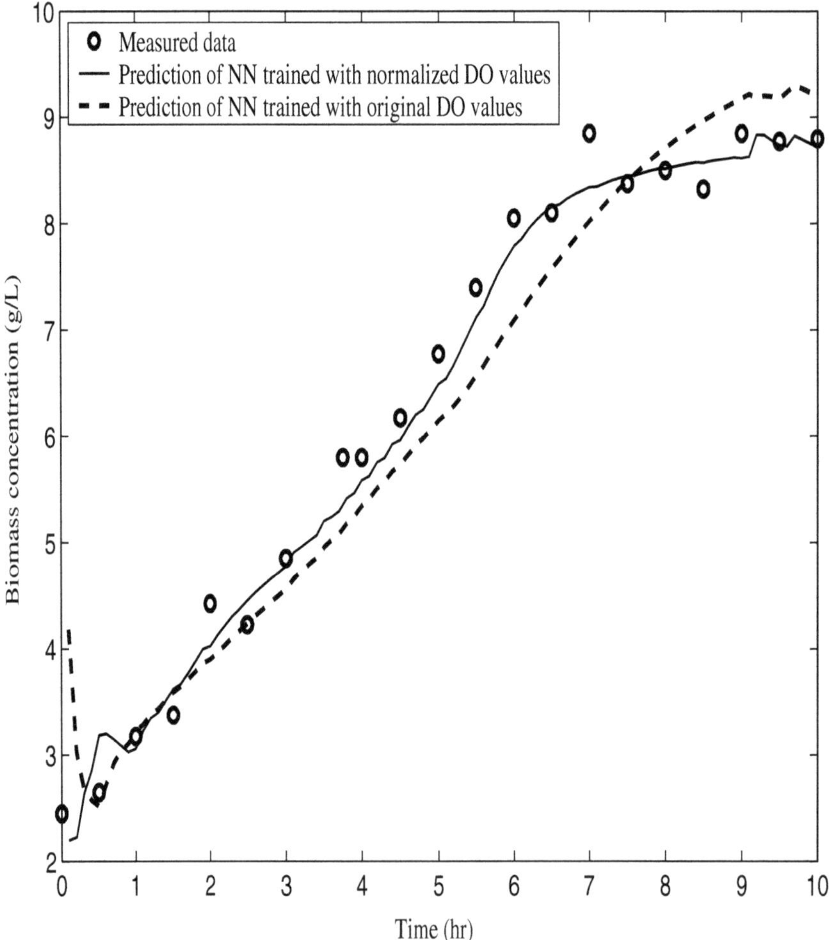

Fig. 6.13. Comparison of biomass prediction based on the neural networks trained with the original DO data and the normalized DO data.

Pre-processing and post-processing of input and output data

For mathematical model identification problems, data processing normally means the filtering of the noisy data. For neural network model identification problems, however, data processing is mainly focused on scaling data so that they can fall into a specific range, which is the most sensitive area of the activation function. In the proposed neural networks, a sigmoid function is used in the activation layers:

$$y = \tanh(\beta x) = \frac{e^{\beta x} - e^{-\beta x}}{e^{\beta x} + e^{-\beta x}} \tag{6.8}$$

where, x is the input to the neuron, y is the output of the neuron, $\beta \in \mathbb{R}$.

The most sensitive input area of the above function is in the range of $[-1, 1]$. The mathematical equation used for input data scaling is given as follows:

$$x_n = 2 \cdot (x - x_{min})/(x_{max} - x_{min}) - 1 \tag{6.9}$$

where, x_n, x, x_{min}, x_{max} are processed value, original value, minimum and maximum value of the original data, respectively.

As the inputs have been transformed into the range of $[-1, 1]$, the output of the trained network will also be in the range $[-1, 1]$. Thus the output data of the neural network have to be converted back to their original units using:

$$y = 0.5 \cdot (y_n) \cdot (y_{max} - y_{min}) + y_{min} \tag{6.10}$$

where, y, y_n, y_{max}, y_{min} are converted value, network output value, maximum value of target data and minimum value of target data, respectively.

Some aspects of data scaling problems were discussed by Koprinkova and Petrova [113]. A major issue was the scaling factor S, which is defined as:

$$S = \frac{R}{B} \tag{6.11}$$

where, B is the highest value of the input, R is the span of the specific range to which the input data will be transformed.

Koprinkova and Petrova found that the smaller S was, the bigger the error would be on the neural network prediction. There was no significant loss of information when $S = 0.009$. The loss of information, however, was unacceptable if S reached the value of 0.00025. In the current study, the highest value of inputs was the upper bound of DO, which was 100. The specific range was $[-1, 1]$, so that the span $R = 2$. Thus the scaling factor $S = 2/100 = 0.02$, which was much higher than 0.009. Furthermore, the scaling factors that were used for all inputs by Koprinkova and Petrova [113] were the same, whereas in this study, different inputs had different scaling factors, such that the scaled values were distributed uniformly through the whole specific range of $[-1, 1]$. This further reduces the loss of information caused by the different input ranges, and it is evident in this study when the above aspects are considered.

Improvement on initial value prediction

As can be seen from Figure 6.13, the initial value prediction was not satisfactory for the trained neural network model. One of the biomass predictions is shown in Figure 6.14. The feed rate profile is f6 as plotted in Figure 6.10. It is obvious that an overshoot occurred at the beginning phase of the prediction. Since the initial culture conditions were the same for all experiments, the most likely reason for this problem was due to the different initial feed rates.

Though an initial prediction problem has been encountered in many neural network modelings, few of them have been solved. An attempt was made by Dacosta et al. [73] to cope with the initial biomass prediction problem when dealing with a radial basis function network model. The different sizes of the initial inoculum were modelled by incorporating a single additional characterization input, which was the initial off-line biomass weight assay. However, the neural network used in this work has a recurrent structure, which makes the proposed method unsuitable in this case.

In order to overcome this problem, a zero-appending method was used. A series of zeros were appended to the beginning of each feed rate. Because the time length must be the same for all input variables, another series of 1s, which equal to the initial values of the reaction volume (1L), was also appended to the data sequence of the fermentation volume. The appended sequences included eight points with six minute intervals. The total time length for appending was 42 minutes. This time length was selected and determined by trial and error. This reflected the unstable period of network prediction. After this period, a stable accurate prediction could be achieved. The result is shown in Figure 6.15.

Results of model prediction

The predictions based on three different models, neural network model I, model II and the mathematical model, were compared as shown in the Figure 6.16. The parameters of the mathematical model, which were identified using the GA, are given in Table 6.1. Among these three biomass prediction curves, the prediction based on neural network model I yielded the best agreement with the experimental data, whereas the mathematical model gave the worst prediction. The overall prediction MSEs for neural models I, II and the mathematical model were 0.1067, 0.3826 and 0.4684 respectively.

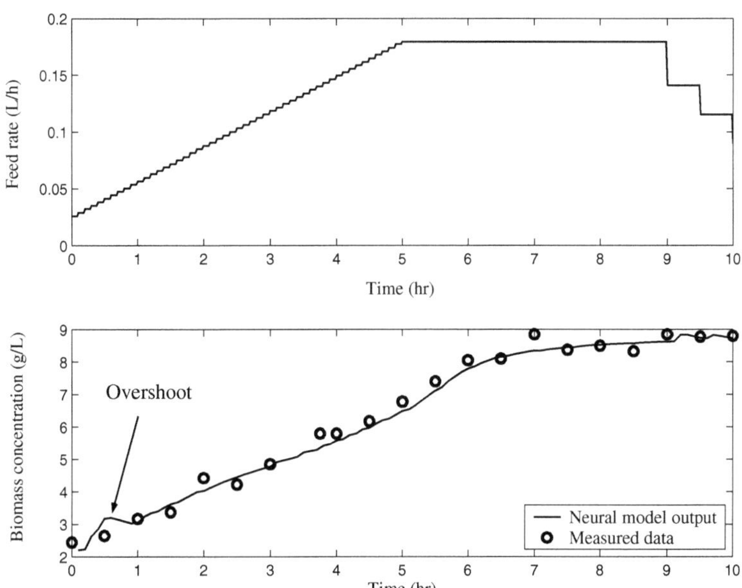

Fig. 6.14. Biomass prediction without using zero appending method.

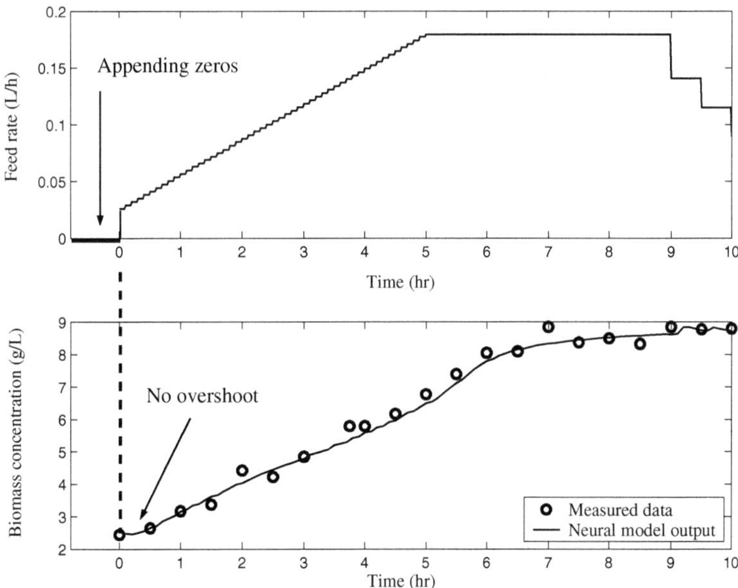

Fig. 6.15. Biomass prediction with zero appending at the beginning of feed rate.

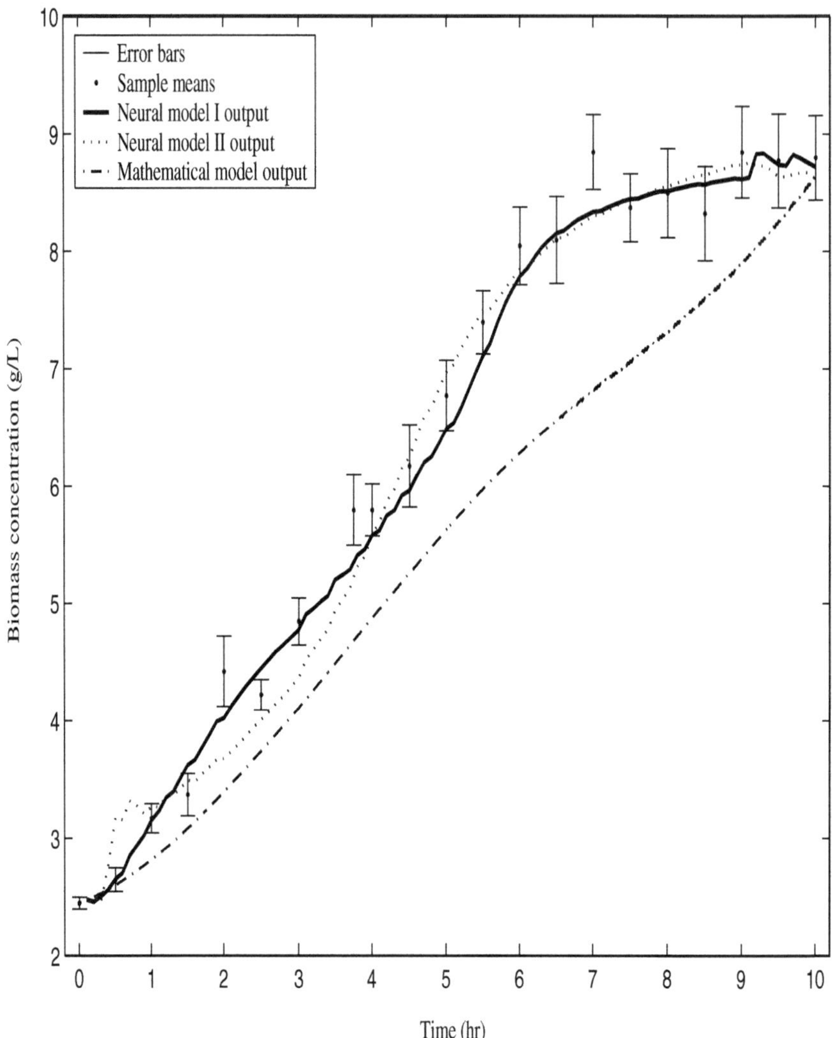

Fig. 6.16. Results of biomass predictions using neural network model I, neural network model II and the mathematical model.

6.5 Conclusions

The Cascade RNN models are proposed in this work to describe a bench-scale fed-batch fermentation of *Saccharomyces cerevisiae*. The nonlinear dynamic behavior of the fermentation process is modelled by the cascade RNN models with internal and external feedback connections. The structures of the models are identified through the training and validating using the data collected from 9 experiments with different feed rate profiles. Data preprocessing methods are used to improve the robustness of the neural network model to match the process' dynamics in the presence of varying initial feed rates. The most accurate biomass prediction is obtained by the DO neural model. The results show that the proposed neural network model has a strong capability of capturing the nonlinear dynamic underlying phenomena of the fed-batch process, provided that sufficient data, measured at appropriate sampling intervals, are available. Results also show that proper data processing and zero-appending methods can further improve the prediction accuracy.

7

Designing and Implementing Optimal Control of Fed-batch Fermentation Processes

This chapter deals with the problem of design and implementation of optimal control for a bench-scale fermentation of *Saccharomyces cerevisiae*. A modified GA is proposed for solving the dynamic constrained optimization problem. The optimal profiles are verified by applying them to the laboratory experiments. Among all 12 runs, the one that is controlled by the optimal feed rate profile based on the DO neural model yields the highest product. The main advantage of the approach is that the optimization can be accomplished without *a priori* knowledge or detailed kinetic models of the processes.

7.1 Definition of an Optimal Feed Rate Profile

The principle of respiratory capacity

The growth of *Saccharomyces cerevisiae* in the fermentation process can be described by the stoichiometries of three pure metabolic routes, namely, oxidative glucose catabolism, reductive (fermentative) glucose catabolism and ethanol utilization [18]. Metabolic pathways that take place during the fermentation are expressed in the following three stoichiometric equations:

$$\text{Oxidation of glucose (R1):} \quad S + a_1 O_2 \overset{r_1}{\to} b_1 X + c_1 CO_2 \tag{7.1}$$

$$\text{Reduction of glucose (R2):} \quad S \overset{r_2}{\to} b_2 X + c_2 CO_2 + d_2 P \tag{7.2}$$

$$\text{Oxidation of ethanol (R3):} \quad P + a_3 O_2 \overset{r_2}{\to} b_3 X + c_3 CO_2 \tag{7.3}$$

where, X, S, P, O_2 and CO_2 are the reaction components, namely, microorganisms, consumed substrate, ethanol, oxygen and carbon dioxide respectively; The parameters a_1, a_3, b_1, b_2, b_3, c_1, c_2, c_3 and d_2 are the stoichiometric coefficients, which are the yields of the three reactions; The reaction rates at which three metabolic pathways take place during the fermentation are r_1, r_2 and r_2 [114].

L. Z. Chen et al.: *Modelling and Optimization of Biotechnological Processes*, Studies in Computational Intelligence (SCI) **15**, 91–108 (2006)
www.springerlink.com

As described in the above equations, according to different fermentation conditions and controls, three different metabolic routes may occur in the growth of microorganisms, which are governed by the respiratory capacity of the cells or so called overflow mechanism. If the substrate flux is low, and there is excess respiratory capacity of the cells, both pathways R1 and R3 are activated, but R1 is observed to have higher priority than R3. Pathway R2 is activated if the substrate flux is high and the respiratory capacity limitations of cells are exceeded.

Overflow metabolism based on bottleneck hypothesis

The limited respiratory capacity can be represented by a bottleneck principle for oxidative substrate utilization [18]. As illustrated in Figure 7.1, three cases (a), (b) and (c) represent the pathways of glucose oxidation, glucose reduction and ethanol oxidation respectively. In case (a), the total amount of substrate can pass the bottleneck, thus the substrate is metabolized purely oxidatively through the pathway R1 (see Equation 7.1). Case (b) represents the substrate flux exceeding the bottleneck of substrate utilization. Part of the substrate that passes the bottleneck is metabolized through the same pathway as in case (a). However, the residual part of the substrate that can not pass the bottleneck is metabolized reductively, and ethanol is formed (see Equation 7.2). Case (c) illustrates the pathway of the oxidation of ethanol, which is described by Equation 7.3.

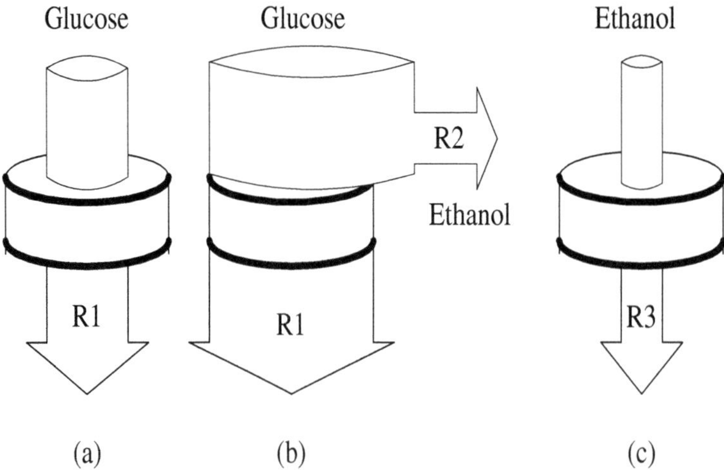

Fig. 7.1. The metabolic pathways represented by the principle of bottleneck.

Restrictions of feed rates in fed-batch fermentations

Even though all three pathways can lead to biomass production, the pathway R2 should be avoided in order to prevent ethanol formation, which results in wasting of the substrate. The pathway R3 takes place only following the occurrence of the pathway R2, and the biomass produced in R3 is very low compared to that produced in the pathway R1. Also, part of the ethanol may be lost due to volatilization. Thus, in order to achieve a high production yield, some control strategies, such as those proposed in [8] and [20], were to assure that the pathway R1 was tightly maintained throughout the fermentation.

For fed-batch fermentation, the substrate concentration inside the reactor can be manipulated by adjusting the feed rate of input substrate. If the residual concentration of substrate in the reaction medium is high, then the substrate flux is high. Conversely, if the residual substrate concentration is low, the substrate flux is low.

A simple illustration of typical stoichiometric constraints on the feed rate during a fed-batch cultivation is shown in Figure 7.2 [115]. In the initial phase of fermentation, the cell density is very low and the feed rate should remain low to avoid overflow metabolism. As the cells grow and cell mass increases, the feed rate can be allowed to increase to meet the nutrient requirement of growing cells. An increased feed rate, however, leads to increased oxygen consumption and eventually the constraint from the limited respiratory capacity may be reached. The feed rate should thus be decreased in order to allow the consumption of residual glucose and ethanol.

Practically, in order to achieve high productivity, a high feed rate is usually needed. However, a bottleneck effect may become significant if the feed rate is too high. To avoid the formation of ethanol, which indicates the occurrence of overflow metabolism, the feed rate should be kept at a sufficiently low level to guarantee only the pathway R1 takes place. In such a case, however, in order to obtain a high productivity, the process may have to be carried out for a long time. This is because there exists an excess respiratory capacity, which is not fully utilized by the cells for growing and reproducing. Moreover, the cells may starve if under-feed happens. On the other hand, both the highest yield and the shortest process time are always desirable for process optimization, since they are of considerable economic importance [56]. However, there is a conflict between these two optimality criteria. A fast production rate usually results in a formation of ethanol, thus decreasing cell mass yield. To avoid ethanol formation, pathway R1 should be maintained throughout the fermentation. This may slow down the whole process. Practically, ethanol formation is always observed during fermentation processes. In past studies, ethanol formation was allowed in some optimal feeding strategies in order to achieve short process times [75]. Produced ethanol was then consumed to produce cell mass at the end of the cultivation. Maintaining the correct balance is therefore of great important to optimizing the production yield.

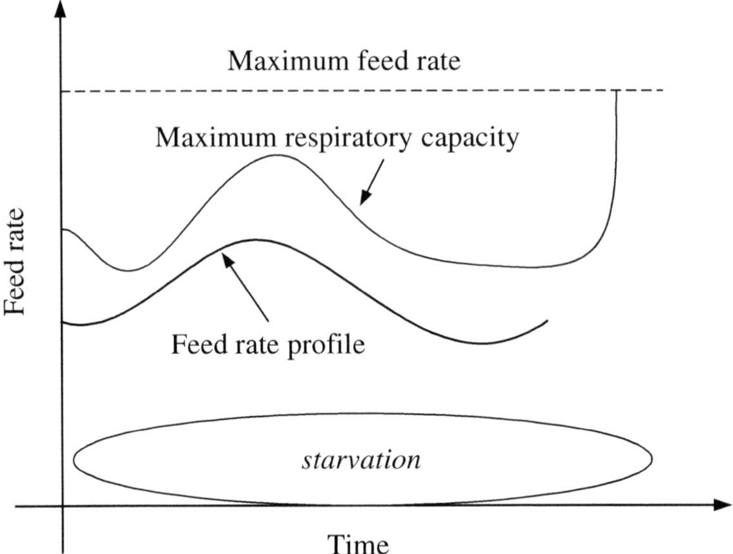

Fig. 7.2. An example of feed rate profile under the constraints of overflow metabolism in a fed-batch fermentation.

Though a number of developed feeding strategies and proposed optimal feed rates can be found in the literature [116, 43], most of them are either set-point or mathematical model-based control. Set-point control or tracking was employed in [8, 20]. In these cases, control strategies were to keep some important variables at their critical values in order to achieve the highest possible productivity. However, a considerable process knowledge is required to make such methods successful. A similar requirement has to be met for mathematical model-based optimization approaches [75, 117]. Recently, a neural network model-based process optimization has drawn considerable attention due to the successful application to highly nonlinear dynamic systems [56] because less *a priori* knowledge is needed to build neural network models than conventional methods. This advantage makes neural network approaches more flexible and transformable from one process to another, thus increasing the development efficiency and benefit-cost ratio.

7.2 Formulation of the Optimization Problem

The performance of fed-batch fermentation of baker's yeast is characterized by the yield obtained at the end of the process. The optimal operation of the fed-batch fermentor can be expressed as: given operational constraints, determine the feeding strategy that optimizes a productivity objective [118].

Thus the optimization problem can be formulated as follows [114]:

$$\max_{F(t)} J = X_{t_f} \times V_{t_f} \tag{7.4}$$

where, t_f is the final process time; X_{t_f} and V_{t_f} are the final biomass concentrations and the final volume, respectively; $F(t)$ is the feed flow rate, which is equally divided into N_{sub} constant control actions:

$$F = [F_1 \quad F_2 \quad \cdots \quad F_{N_{sub}}]^T \tag{7.5}$$

where, N_{sub} is the number of intervals of the feed rate profile.

The optimization is subject to the following constraints:

$$\begin{aligned} 0 \leq \ F(t) \ &\leq 0.2988 \ \text{L/h} \\ V(t_f) &= V_{max} = 2.5 \ \text{L} \\ t_f \quad &= 8 \ \text{hrs} \end{aligned} \tag{7.6}$$

The initial conditions and the glucose concentration of feeding solution are given in Table 6.1 as shown in Chapter 6.

7.3 Optimization Procedure

In order to avoid the optimization procedure being stuck in local maximum, a simple strategy is used in the search procedure. The feed rate profile is firstly divided into four piecewise of control variables with equal time length, which are used as the variables of the GA to find the optimal solution. After the convergency of the GA, the final four piecewise feed rate are then divided into eight equal lengths of sub-feed rates. The GA is then run again, the feed rate is then divided again, and so on. It terminates when the improvement on performance index is less than a predefined value over a certain number of iterations or when a predefined maximum number of intervals, N_{max}, has been reached. The flowchart of this strategy is illustrated in Figure 7.3.

When the subinterval of the process horizon is decreased (i.e., the subdivision of control actions is increased), the final (best) population obtained from the preceding run of the GA should be divided according to the new subdivisions. A subdivisional operation is thus required to divide the population into smaller subintervals. The time steps in the evaluation function should be also updated corresponding to the change of time intervals in the population.

The constraint on the final volume given in Equation 7.6 is implemented as a penalty function. Usually, a validity checking procedure, through which each candidate solution has to pass, is adopted to isolate the solution that does not hold the constraint. If the final volume produced by a feed rate solution is not two liters, the fitness value of this solution is set to be zero, which means this individual has less chance to survive. Otherwise, the fitness value of the candidate solution is evaluated using the objective function given

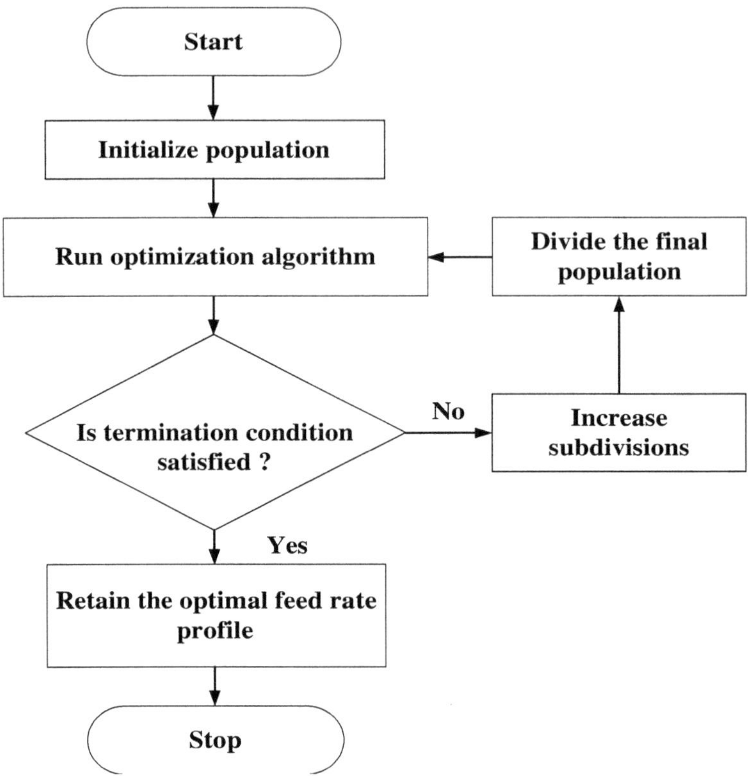

Fig. 7.3. Flowchart of the optimization strategy.

in Equation 7.4. However, an invalid feed rate solution may possess some properties which are close to those in the optimal solution, especially for those with high fitness values (before set to be zero) and final volume that is less than two liters. The fitness values being simply set to zero will deprive them of the chance to evolve into the optimum. On the other hand, a slow convergency may be observed due to the fitness values of a large number of candidate solutions being set to be zero.

Instead of giving zero to the fitness value, a penalty form of objective function was employed to calculate the fitness value for the solution that violated the constraint [45]. A penalized fitness value was assigned to the solution depending on how far it deviated from the limit:

$$\begin{aligned} J &= X_{t_f} \times V_{t_f} - r \cdot [V_{max} - V(t_f)]^2 \qquad && \text{If } V(t_f) > V_{max} \\ J &= X_{t_f} \times V_{t_f} \qquad && \text{If } V(t_f) \leq V_{max} \end{aligned} \qquad (7.7)$$

where r is the penalty coefficient, which was assigned to the value of 100.

However, the constraint violation still remains after applying the penalty form of objective function. This may also reduce the convergence rate as it takes a long time to find a solution with the highest fitness value as well as holding the constraint.

To solve the above problem, a new method proposed in this study is to amend or repair the conflicting candidate solutions. There are three cases for each candidate solution being evaluated. In case one, the final volume is the same as the constraint value, so no modifications are needed for both the solution and its fitness value. In case two, the final volume produced by the solution is higher than the constraint value. The new fitness of this candidate is simply set to zero because it is usually not a promising solution. Even though a high fitness value appears, it is normally caused by a high feed rate rather than an optimal feed rate. In case three, the final volume produced is lower than the constraint value. This solution is rectified to meet the constraint requirement by adding the deficiency uniformly in order to produce the final volume of V_{max}. In other words, the feed rate trajectory of this solution is moved up in parallel to produce the required final volume, where the shape of the trajectory is preserved. The new procedure can be summarized as follows:

1. Evaluate the fitness value J according to Equation 7.4 for the candidate solution F_i, $i = 0, 1, \cdots, n$, calculate the final volume V that produced by F_i.
   a) If $V = V(t_f)$, $F_{inew} = F_i$, $J_{new} = J$.
   b) If $V > V(t_f)$, $F_{inew} = F_i$, $J_{new} = 0$.
   c) If $V < V(t_f)$, $F_{inew} = F_i + (V(t_f) - V)/t_f$, re-evaluate the fitness value J_{new} according to Equation 7.4.
2. Run the GA using the new fitness value J_{new}.
3. Repeat steps 1-2, until a termination criterion is reached.

7.4 Optimization Results and Discussion

Constraint handling

The constraint handling method proposed in this work was tested in comparison with the conventional one. Figure 7.4 shows the convergence profiles of the GA using two different constraint handling methods. The best performance indices of each generation were plotted against the generation count. Neural network model I was employed in the tests and a fixed number of intervals, eight, was used to divide the feed rate control profiles.

It can be seen from Figure 7.4 that both indices climbed up to 24 at the starting points of the optimization runs. However, from the range of 24.5-26, a rapid increase of the fitness value was observed for the proposed constraint handling method. The maximum performance index was achieved within 50 generations, while the conventional method took 250 generations to reached the maximum performance index. Several temporary stallings appeared in the performance index of the conventional method. This was due to the algorithm being stuck in some local maxima because the fitness values of the solutions

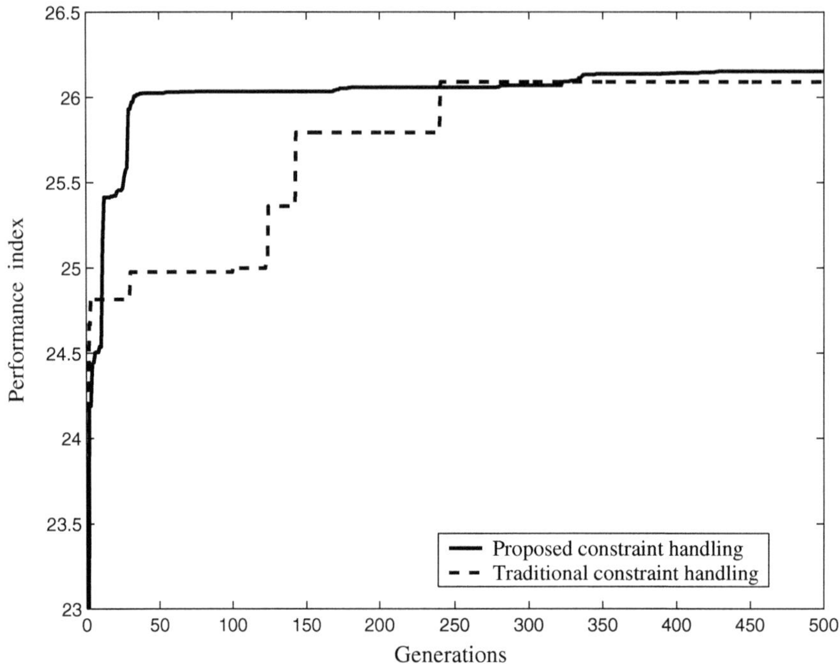

Fig. 7.4. Convergence profiles for different constraint handling methods.

that violated to the constraint were simply set to zero. With the strong tendency towards the global optimum, the GA was able to continuously escape from such local maxima and gradually converge to the optimal value of performance index. However, the proposed method is superior to the conventional method with the advantage of fast convergence speed.

Development of the optimal feed rate profiles

An investigation into the optimization by using different time subintervals for dividing the feed rate control actions was conducted in order to develop the suitable optimization strategy. Figure 7.5 shows the best performance indices for optimizations with different number of intervals: 4, 8, 16 and 80.

As discussed above, with the stochastic nature, the GA can eventually reach the similar optimum points for different number of intervals with different convergence rates. It can be seen from Figure 7.5 that the smaller the number of intervals, the faster the convergence rate. The algorithm took about 50 generations to converge when the feed rate was divided into four piecewise control actions, while it took about 1000 generations to reach the optimal value of index when the feed rate was divided into 80 sub-feed actions. On the other hand, because the value of feed rate within the subinterval is fixed, a

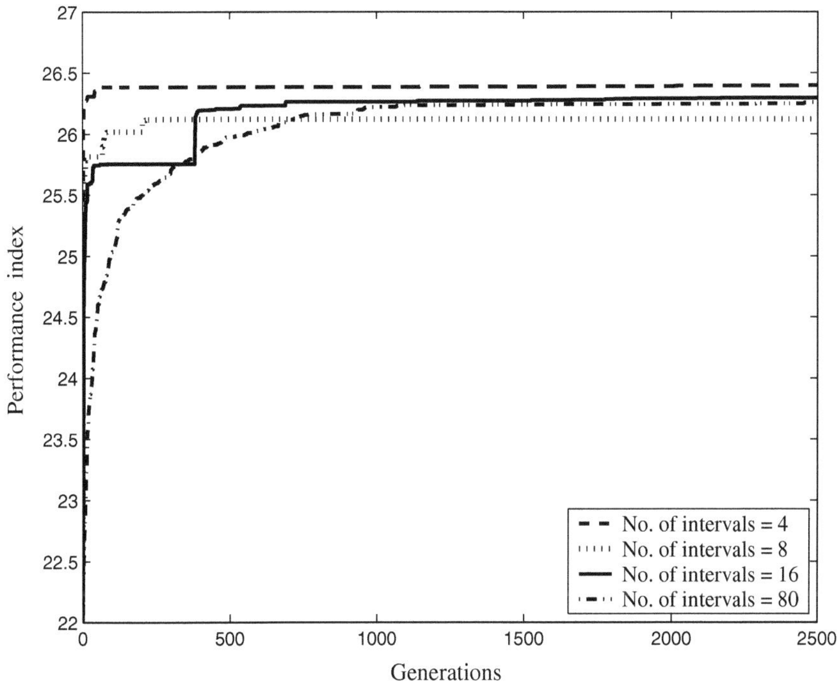

Fig. 7.5. Convergence profiles for different number of intervals.

smaller number of intervals means a longer time for each interval, and it gives less flexibility to the feed rate profile. As discussed by Na et al. [75], the necessary conditions for optimality at different subintervals rather than a given subinterval on the profile can not be fulfilled. In this sense, a feed rate profile that is obtained using a small number of intervals (e.g. four) may be viewed in principle as a suboptimal result, although it may produce a performance index value which is very close to others. In fact, as seen in Figure 7.5, the performance index of the feed rate with four intervals is higher than those with higher number of intervals. However, the index value for four intervals could hardly improve once it converges from the generation number of 50, whereas the indices for 16 and 80 intervals evolve slowly from starting point up to the end of the optimization.

A novel feature of the proposed optimization strategy is the incremental interval number technique. In order to achieve faster convergence as well as a higher performance index, the number of intervals was initially chosen as four, then was increased to eight, 16, and 80 during the progress of the optimization procedure. Figure 7.6 shows the evolvement of the performance index when using the proposed optimization strategy. Due to the fast convergence, the four intervals was used in the first 100 generations. The end population was then transferred into the population with the feed rate represented by eight piecewise control actions, and the GA was run on the new population for 500 generations. The same procedure was carried out until the last run of the GA with 80 intervals. The fast convergence was achieve by the feed rate with four intervals, and a higher index was obtained by further dividing the feed rate profile based on the result of the previous stage. A relatively big improvement was observed for the feed rate with eight intervals. As the number of intervals increased, the improvements became smaller.

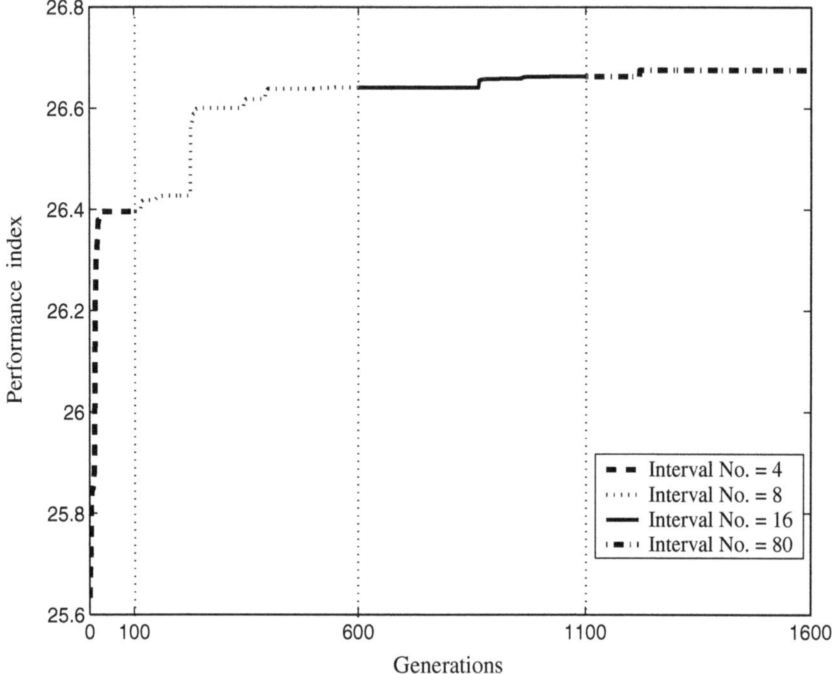

Fig. 7.6. Performance index profiles for the proposed optimization strategy.

Comparisons of computation times for different number of intervals are given in Table 7.1. The optimizations were performed on a IBM compatible computer with an Intel Pentium II Celeron 633 MHz processor. The population size was chosen as 250.

The top half of Table 7.1 shows the running times of fixed number of intervals, i.e., the number of intervals was unchanged from the beginning to the end of the optimization. The bottom half of the table shows the computation times spent with the incremental number of intervals, i.e., the number of intervals was increased from four to 80 during the course of the optimization. It can be seen from the table that the times spent were the same for the optimizations when the number of intervals equalled four. However, from intervals eight to 80, the computation times of optimization with an incremental number of intervals were shorter than those with fixed number of intervals. This is due to the end population of a previous stage is subsequently converted into the initial population of the next stage when the subdivision was increased. Thus the times for initialization were saved.

Table 7.1. The computation times for fixed number of intervals and incremental number of intervals.

Intervals	Generations	Computation Time (hrs)
4 (fixed)	200	2.27
8 (fixed)	200	2.38
16 (fixed)	200	2.60
80 (fixed)	200	2.82
Fixed No. of intervals total	800	10.07
4 (start)	200	2.27
$\downarrow$ 8	200	2.01
$\downarrow$ 16	200	2.10
$\downarrow$ 80 (end)	200	2.16
Incremental No. of intervals total	800	8.54

The development of optimal feed rate profile using the optimization strategy is shown in Figure 7.7(a)-(d). A significant change of feed rate profile appeared from number of intervals four to eight. From number eight to 80, however, there was no big alteration in the shape of the profile. It was only "fine tuned" by the optimization procedure when it was further subdivided.

A large number and high frequency of fluctuations appeared on the the feed rate profile with 80 intervals as shown in Figure 7.7(a). This makes the profile impractical to implement using the laboratory controller. Moreover, due to the continuous behavior of the fermentation process, a highly fluctuating feed rate may cause unexpected disturbances to the system [80]. Practically, further increase in subdivision for optimization might lead to even worse results [45]. In the experiments of this study, 16 was chosen as the final number of intervals to divide the entire feed rate into equal length of sub-feed rates. This selection is also a compromise between running time and the performance index. Higher subdivision may result in higher yield but is more time consuming.

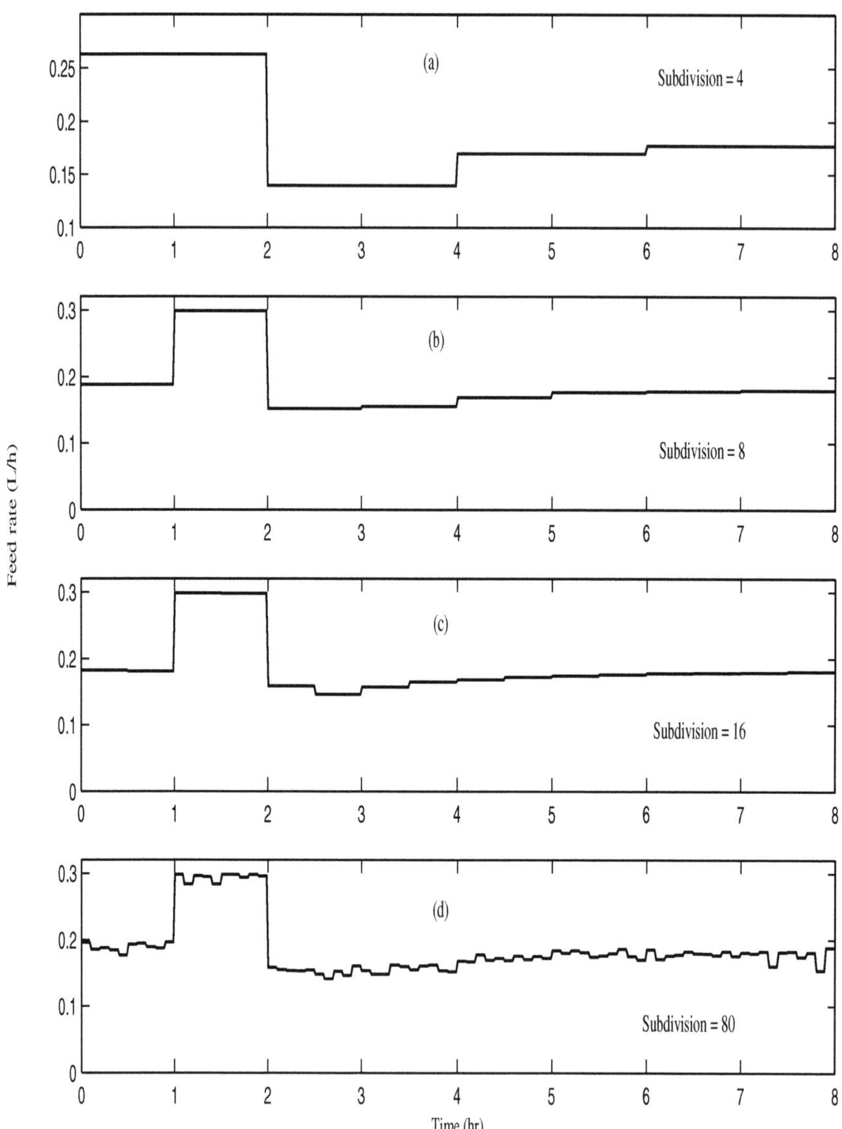

Fig. 7.7. Evolution of the optimal feed rate profile using the proposed optimization strategy.

Implementation of the optimal controls

The optimal control based on neural model I

Three experimental runs controlled by different optimal feed rate profiles were carried out using BioFlo 3000 bench-scale fermentors. The detailed experimental setup and procedures are described in Chapter 6. The result of optimization based on the proposed neural network model I and experimental validation are given in Figure 7.8. The optimal feed rate profile is plotted in Figure 7.8(a). This shows the substrate was fed into the fermentor at a middle level of feed rate (0.18 L/h) during the first hour. The feed rate then rose rapidly to its maximum value (0.2988 L/h) to allow a fast feeding. In the second hour, the feed rate was decreased from the maximum value to a lower value (0.16 L/h). After half an hour, it was further decreased to 0.15 L/h, which was the minimum feed rate during the whole fermentation process. In the third hour, the feed rate was increased gradually with a small extent (< 0.01 L/h) every half an hour until the end of the process.

This feed rate profile conforms to the overflow metabolism of the growth of *Saccharomyces cerevisiae*. At the beginning of the fermentation, the seed of microorganisms is just inoculated into the reactor. The feed rate is not permitted to be high in order to allow the cells to adapt to the new condition of growth. A gentle middle level of feed rate allows the cells to get into the exponential growth phase as quickly as possible, without the overflow metabolism taking place. As the cells are brought to the active state, a high feed rate is required to supply sufficient nutrients to the cells and allows the cells to grow adequately. However, as explained in Section 7.1, a continuous high feed rate leads to increased oxygen consumption and overflow metabolic pathways. Thus, a switch to a low feed rate is necessary to avoid the bottleneck effect and in the same time, to allow the consumption of residual glucose and ethanol.

The biomass concentration produced by the optimal feed rate is shown in Figure 7.8(b). The final biomass concentration of the experimental result was 11.02 g/L, which is the highest among all of those obtained from experimental runs in this research, which are summarized in Table 7.2. From Figure 7.8(b)-(c), one can see a close prediction of biomass growth under optimal condition was also achieved. The final biomass predicted was 10.6698 g/L, which was in a good agreement with the experimental result (11.02 g/L). The prediction percentage error was less than 10% during the whole fermentation period.

The optimal control based on neural model II

The optimization result based on neural network model II is given in Figure 7.9(a)-(c). As shown in Figure 7.9(a), the feeding started with a very high feed rate (0.24 L/h) and lasted for four hours. A low feed rate was then used from the fourth to fifth hour followed by another high feed rate until the end of the fermentation. Obviously, this feed rate profile was in conflict with the

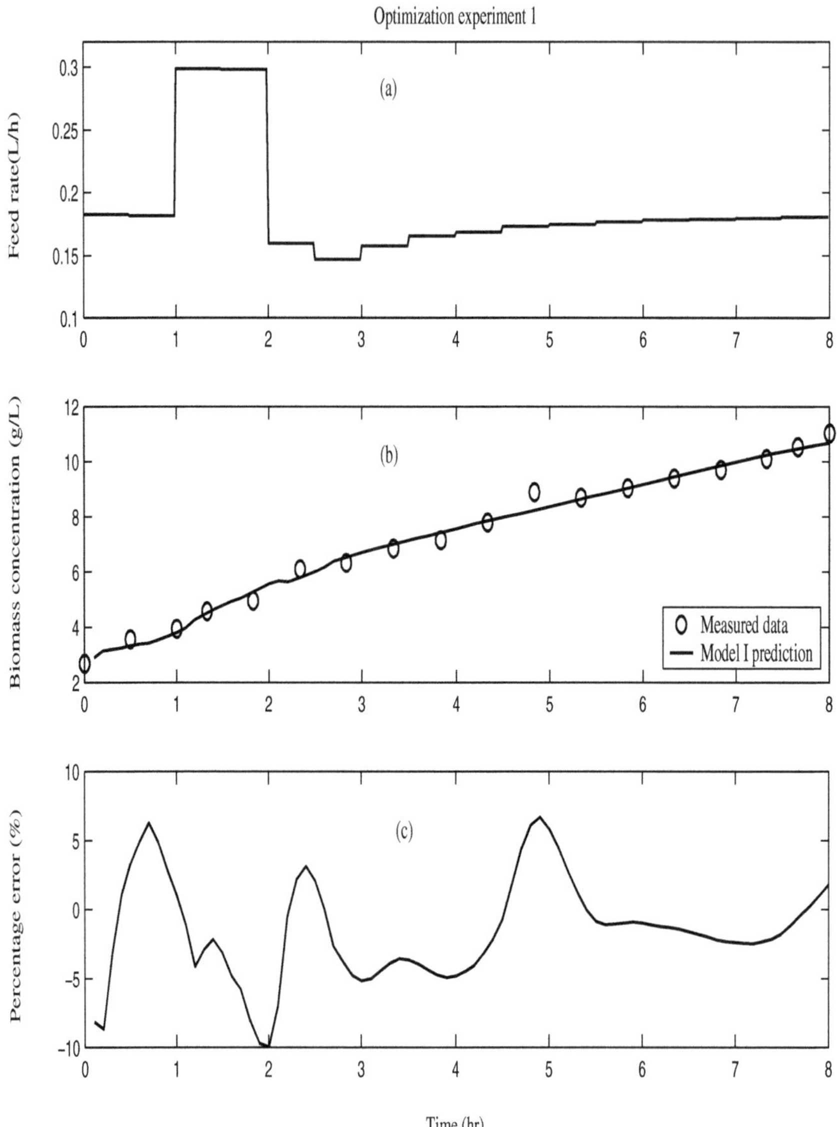

Fig. 7.8. Optimization result based on the cascade recurrent network model I

overflow metabolism and the neural network model II failed to predict the optimal trajectory. The produced biomass concentration was the lowest one among the three optimization runs as can be seen from Table 7.2. The main reason for this may be due to the neural network being trained with insufficient glucose data, which were measured at a long sampling time (30 minutes) in this work. However, as can be found from the error plot in Figure 7.9(c), the neural network surprisingly estimated the biomass concentration with similar accuracy to that of neural model I, even though it is not the optimal trajectory. The most likely explanation for this result is that this neural model can predict well in the "experimental space", which is the space spanned by the measurement data used for training the network. However, its extrapolation property is poor [119]. It is expected that similar results can be achieved as those obtained from neural model I, if an on-line glucose sensor, which can detect glucose concentration more frequently (e.g., every five minutes), is available.

The optimal control based on the mathematical model

For a comparison, the results of optimization based on the unstructured mathematical model (Equations 6.4 - 6.5) is presented in Figure 7.10(a)-(c). The feed rate profile, as plotted in Figure 7.10(a), has a similar pattern as that shown in Figure 7.8(a). The difference between these two profiles is that the highest feed rate (about 0.25 L/h) lasted for two hours, while Figure 7.8(a) shows that the highest feed rate (0.2988 L/h) lasted for just one hour. The final biomass produced was 9.53 g/L as shown in Figure 7.8(b), which is less than that obtained using neural model I, but higher than that obtained from neural model II. From the error plot in Figure 7.8(c), it is found that the prediction accuracy is not as good as the neural models. However, a very accurate prediction was made on the final biomass concentration. From the above results, it is obvious that the mathematical model is less accurate in prediction during the initial and the middle phase of the fed-batch fermentation process, but it has a good extrapolation ability to predict the final biomass concentration [108]. The performance of the mathematical model for optimization is better than that of the neural model trained with insufficient information, but is worse than that of the neural model trained with adequate data sets.

The final biomass concentrations and total reaction times of the experiments that were carried out in this research are summarized in Table 7.2. Run 1 to run 9 are correspondingly the experiments that used the feed rate f1 to f9 shown in Figure 6.10. Run op1 to op3 were the optimization experiments one to three as shown in Figure 7.8, 7.9 and 7.10, respectively. Table 7.2 shows that the highest biomass concentration and shortest reaction time was achieved in the run op1, which is the optimization experiment based on the neural network model I.

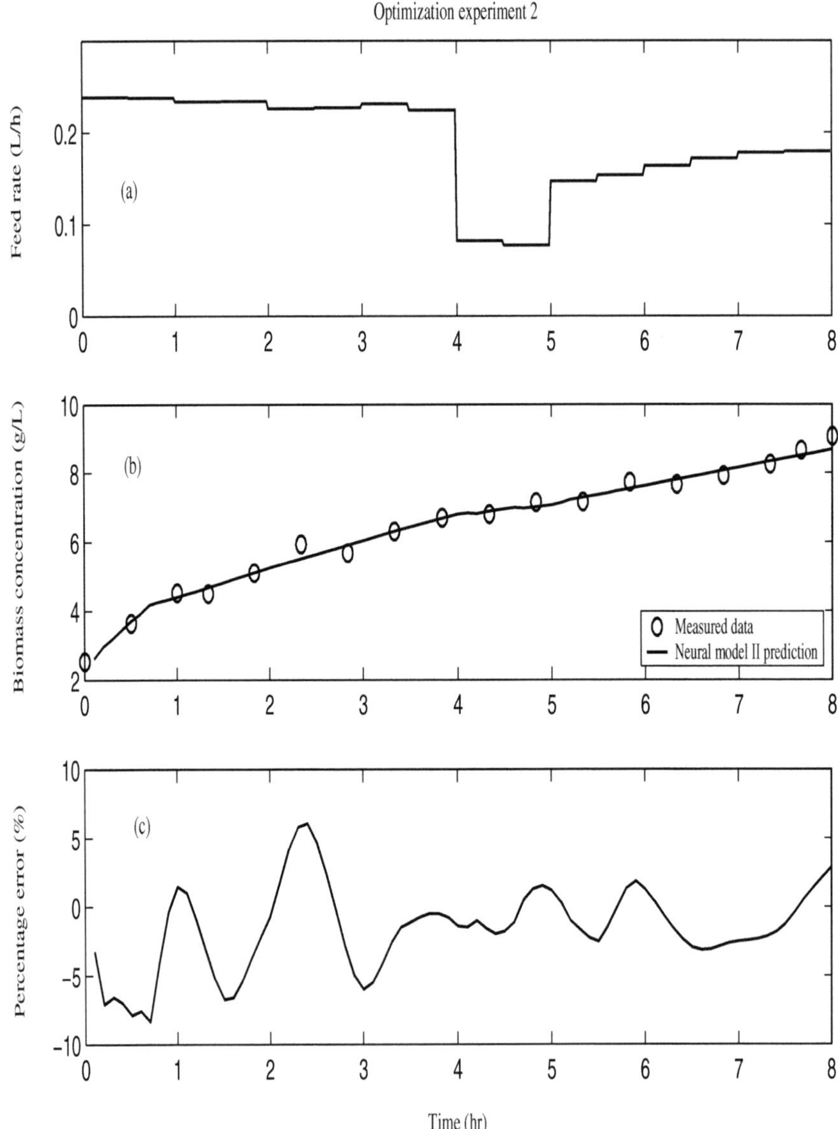

Fig. 7.9. Optimization result based on the cascade recurrent network model II.

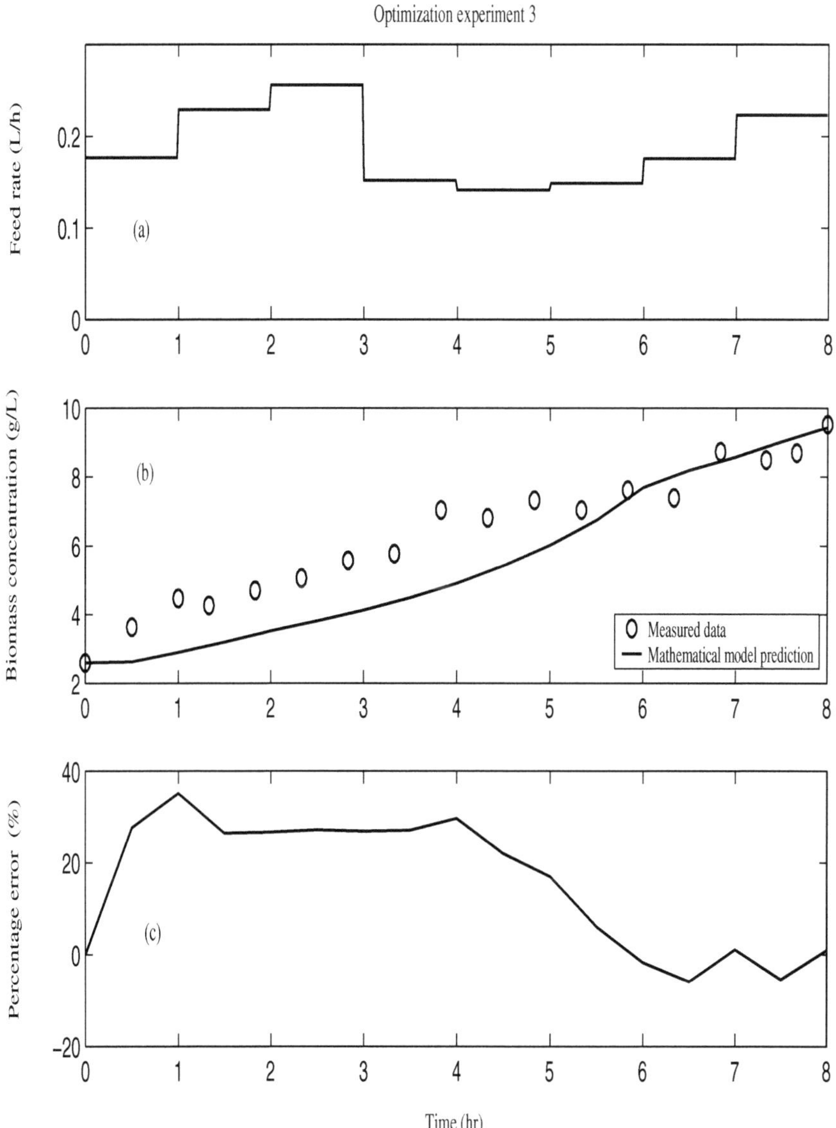

Fig. 7.10. Optimization result based on the mathematical model.

Table 7.2. The measured and predicted final biomass concentrations and total reaction times for all experiments that have been carried out in this study.

Run	Total time (hr)	Final biomass (g/L)	
		Predicted	Measured
1	12.5	-	8.45
2	12.5	-	7.6
3	12.5	-	9.2
4	12.5	-	9.65
5	12.5	-	8.5
6	12.5	-	9.5
7	12.5	-	6.575
8	12.5	-	8.0
9	12.5	-	9.65
op1	8	10.67	11.02
op2	8	8.68	9.05
op3	8	9.44	9.53

7.5 Conclusions

The design and experimental implemention of optimal feed rate profiles is described in this chapter. The modified GA is presented for solving the dynamic constraint optimization problem. The fast convergence as well as the global solution are achieved by the novel constraint handling method and incremental subdividing of the feed rate profile. The optimal profiles are verified by applying them to laboratory scale experiments. Among all 12 runs, the one controlled by the optimal feed rate profile based on the DO neural model gives the highest biomass concentration at the end of the fermentation process. The main advantage of the approach proposed in this work is that the optimization can be accomplished without *a priori* knowledge or detailed kinetic models of the processes. Owing to the data-driven nature of neural networks and the stochastic search mechanism of the GA, the approach can be readily adopted for other dynamic optimization problems such as determining optimal initial conditions or temperature trajectories for batch or fed-batch reactors.

8

Conclusions

8.1 General Conclusions

In this book, a number of results related to monitoring, modelling and optimization of fed-batch fermentation processes are presented. The study focuses on AI approaches, in particular, RNNs and GAs. These two techniques can be used either separately or together to fulfill various goals in the research. The great advantages that are offered by these approaches are the flexible implementation, fast prototype development and high benefit/cost ratio. Their applications to biotechnology process control provide a new inexpensive, yet effective way to improve the production yield and reduce the environmental impact.

A comparison of GAs and DP has demonstrated that GAs are superior to DP for optimization fed-batch fermentation processes. An on-line identification and optimization method based on a series of real-valued GAs was successfully applied to estimate the parameters of a seventh order system and to maximize the final concentration of hybridoma cells in a fed-batch culture. In the first two days of the fermentation, the system parameters were found using the GA based on the measured data. Then the optimal feed rate control profiles were determined using the predicted model. In the last eight days of fermentation, the bioreactor was driven under the control of optimal feed flow rates and reached a final MAb concentration of 193.1 mg/L and a final volume of $2L$ at the end of the fermentation. This result is only 2% less than the best result (196.27 mg/L) obtained in the case which all the parameters are assumed to be known.

The suitability of using a RNN model for on-line biomass estimation in fermentation processes has been investigated. Through simulations, an appropriate neural network topology is selected. This selected neural network topology is further tested experimentally. From the experimental results, the proposed softsensor has shown itself be able to predict the biomass concentration with an RMSP error of 10.3%. The proposed softsensor provides a powerful tool for measuring the biomass on-line.

L. Z. Chen et al.: *Modelling and Optimization of Biotechnological Processes*, Studies in Computational Intelligence (SCI) **15**, 109–111 (2006)
www.springerlink.com

A cascade RNN model proposed in this work has proved capable of capturing the dynamic nonlinear underlying phenomena contained in the training data set and can be used as the model of the bioprocess for optimization purpose. The structure of the neural network model is selected using validation and testing methods. A modified GA is presented for solving the optimization problem with a strong capability of producing smooth feed rate profiles. The results of optimal feeding trajectories obtained based both on the mechanistic model and the neural network model have demonstrated that the cascade recurrent neural model is competent in finding the optimal feed rate profiles. The proposed approach can partly eliminate the difficulties of having to specify completely the structure and parameters of a bioprocess model.

Finally, the design and implementation of optimal control of bench-scale fed-batch fermentation processes using cascade RNNs and GAs are presented. The neural network that is proposed in the work has a strong capability of capturing the nonlinear dynamic relationships between input-output data pairs, provided that sufficient data that are measured at appropriate sampling intervals are available. It has also shown that proper data processing and zero-appending methods can further improve the prediction accuracy. GAs have been used for solving the dynamic constraint optimization problem. The fast convergence as well as global solution are achieved by the novel constraint handling technique and the incremental feed subdivision strategy. Among all 12 experiments, the one controlled by the optimal feed rate profile based on the DO neural model yields the highest product. The main advantage of the approach is that the optimization can be accomplished without *a priori* knowledge or detailed kinetic models of the processes. Owing to the data-driven nature of neural networks and the stochastic search mechanism of GAs, the approach can be readily adopted for other dynamic optimization problems such as determining optimal initial conditions or temperature trajectories for batch or fed-batch reactors.

8.2 Suggestions for Future Research

Investigations presented in this book have opened several key areas that the author would like to suggest for future studies.

- Combination of problem-specific process knowledge and RNNs can be considered to enhance the robustness and extrapolability of the fed-batch fermentation model. However, the development cost may increase.
- Combination of conventional mathematical optimization schemes with the GA should further improve the optimality of the optimal feed rate profiles.
- Online adaptation or tuning of the models and the optimal feed rate profiles are required to produce more reliable and repeatable results, especially when the process time is increased.
- Optimal experimental design can be used to increase the span of the space that is covered by the experimental database.

A

A Model of Fed-batch Culture of Hybridoma Cells

A mathematical model for fed-batch culture of hybridoma cells [24] has been employed for generating simulation data in this study. The model is a seventh-order nonlinear model where both glucose and glutamine concentrations are used to describe the specific growth rate, μ. The cell death rate, k_d, is governed by lactate, ammonia and glutamine concentrations. The specific MAb production rate, q_{MAb}, is estimated using a variable yield coefficient model related to the physiological state of the culture through the specific growth rate. The mass balance equations for the system in fed-batch mode are:

$$
\begin{aligned}
\frac{dX_v}{dt} &= (\mu - k_d)X_v - \frac{F}{V}X_v \\
\frac{dGlc}{dt} &= (Glc_{in} - Glc)\frac{F}{V} - q_{glc}X_v \\
\frac{dGln}{dt} &= (Gln_{in} - Gln)\frac{F}{V} - q_{gln}X_v \\
\frac{dLac}{dt} &= q_{lac}X_v - \frac{F}{V}Lac \\
\frac{dAmm}{dt} &= q_{amm}X_v - \frac{F}{V}Amm \\
\frac{dMAb}{dt} &= q_{MAb}X_v - \frac{F}{V}MAb \\
\frac{dV}{dt} &= F
\end{aligned}
\tag{A.1}
$$

with the following kinetic expressions:

$$
\begin{aligned}
\mu &= \mu_{max}\left[\frac{Glc}{K_{glc}+Glc}\right]\left[\frac{Gln}{K_{gln}+Gln}\right] \\
k_d &= k_{dmax}(\mu_{max} - k_{dlac}Lac)^{-1}(\mu_{max} - k_{damm}Amm)^{-1}\frac{k_{dgln}}{k_{dgln}+Gln} \\
q_{gln} &= \frac{\mu}{Y_{xv/gln}} \\
q_{glc} &= \frac{\mu}{Y_{xv/glc}} + m_{glc}\left[\frac{Glc}{k_{mglc}+Glc}\right] \\
q_{lac} &= Y_{lac/glc}\,q_{glc} \\
q_{amm} &= Y_{amm/gln}\,q_{gln} \\
q_{MAb} &= \alpha'\mu + \beta \qquad \text{where} \qquad \alpha' = \left[\frac{\alpha_0}{k_\mu + \mu}\right]
\end{aligned}
\tag{A.2}
$$

where X_v, Glc, Gln, Lac, Amm and MAb are respectively the concentrations in viable cells, glucose, glutamine, lactate, ammonia and monoclonal antibodies; V is the fermentor volume and F the volumetric feed rate; Glc_{in} and Gln_{in} are the concentrations of glucose and glutamine in the feed stream,

L. Z. Chen et al.: *Modelling and Optimization of Biotechnological Processes*, Studies in Computational Intelligence (SCI) **15**, 111–112 (2006)
www.springerlink.com

respectively; q_{glc}, q_{gln}, q_{lac}, q_{amm} and q_{MAb} are the specific rates; $Y_{xv/gln}$, $Y_{xv/glc}$ and $Y_{lac/glc}$ are yield coefficients. The parameter values are tabulated in Table A.1.

Table A.1. The parameter values of the kinetic model

Parameters	Values
μ_{max}	$1.09d^{-1}$
k_{dmax}	$0.69d^{-1}$
$Y_{xv/glc}$	$1.09 \times 10^8 cells/mmol$
$Y_{xv/gln}$	$3.8 \times 10^8 cells/mmol$
m_{glc}	$0.17mmol \cdot 10^{-8}cells \cdot d^{-1}$
k_{mglc}	$19.0mM$
K_{glc}	$1.0mM$
K_{gln}	$0.3mM$
α_0	$2.57mg \cdot 10^{-8}cells \cdot d^{-1}$
K_μ	$0.02d^{-1}$
β	$0.35mg \cdot 10^{-8}cells \cdot d^{-1}$
k_{dlac}	$0.01d^{-1}mM^{-1}$
k_{damm}	$0.06d^{-1}mM^{-1}$
k_{dgln}	$0.02mM$
$Y_{lac/glc}$	$1.8mmol/mmol$
$Y_{amm/gln}$	$0.85mmol/mmol$

The multi-feed case, which involves two separate feeds F_1 and F_2 for glucose and glutamine respectively, is reformulated as follows:

$$
\begin{aligned}
\frac{dX_v}{dt} &= (\mu - k_d)X_v - \frac{F_1+F_2}{V}X_v \\
\frac{dGlc}{dt} &= (Glc_{in} - Glc)\frac{F_1+F_2}{V} - q_{glc}X_v \\
\frac{dGln}{dt} &= (Gln_{in} - Gln)\frac{F_1+F_2}{V} - q_{gln}X_v \\
\frac{dLac}{dt} &= q_{lac}X_v - \frac{F_1+F_2}{V}Lac \\
\frac{dAmm}{dt} &= q_{amm}X_v - \frac{F_1+F_2}{V}Amm \\
\frac{dMAb}{dt} &= q_{MAb}X_v - \frac{F_1+F_2}{V}MAb \\
\frac{dV}{dt} &= F_1 + F_2
\end{aligned}
\tag{A.3}
$$

The following initial culture conditions and feed concentrations are used in the work:

$$
\begin{aligned}
X_v(0) &= 2.0 \times 10^8 cells/L \\
Glc(0) &= 25mM \\
Gln(0) &= 4mM \\
Lac(0) &= Amm(0) = MAb(0) = 0 \\
Clc_{in} &= 25mM \\
Gln_{in} &= 4mM \\
V(0) &= 0.79L
\end{aligned}
\tag{A.4}
$$

The above mathematical models and initial conditions have been used to generate a 'reality' for testing the schemes proposed in the work.

B

An Industrial Baker's Yeast Fermentation Model

A mathematical model of an industry fed-batch fermentation process, which was given in [19], is used to describe the system. The kinetics of yeast metabolism that is considered in the model is based on the bottleneck hypothesis [18]. The model is governed by a set of differential equations derived from mass balances in the system. It comprises the following equations:

Balance equations:

$$\frac{d(V \cdot C_s)}{dt} = F \cdot S_0 - \left(\frac{\mu}{Y_{x/s}^{ox}} + \frac{Q_{e,pr}}{Y_{e/s}} + m\right) \cdot V \cdot X \tag{B.1}$$

$$\frac{d(V \cdot C_o)}{dt} = -Q_o \cdot V \cdot X + k_L a_o \cdot (C_o^* - C_o) \cdot V \tag{B.2}$$

$$\frac{d(V \cdot C_c)}{dt} = Q_c \cdot V \cdot X + k_L a_c \cdot (C_c^* - C_c) \cdot V \tag{B.3}$$

$$\frac{d(V \cdot C_e)}{dt} = (Q_{e,pr} - Q_{e,ox}) \cdot V \cdot X \tag{B.4}$$

$$\frac{d(V \cdot X)}{dt} = \mu \cdot V \cdot X \tag{B.5}$$

$$\frac{dV}{dt} = F \tag{B.6}$$

where, C_s, C_o, C_c, C_e, X, and V are state variables which denote concentrations of glucose, dissolved oxygen, carbon dioxide, ethanol, and biomass, respectively; V is the liquid volume of the fermentation; F is the feed rate which is the input of the system; m is the glucose consumption rate for the maintenance energy; $Y_{e/s}$ and $Y_{x/s}^{ox}$ are yield coefficients; $k_L a_o$ and $k_L a_c$ are volumetric mass transfer coefficients; S_0 is the concentration of feed.

Glucose uptake rate:

$$Q_s = Q_{s,max} \frac{C_s}{K_s + C_s} \tag{B.7}$$

Oxidation capacity:

L. Z. Chen et al.: *Modelling and Optimization of Biotechnological Processes*, Studies in Computational Intelligence (SCI) **15**, 113–114 (2006)
www.springerlink.com

$$Q_{o,lim} = Q_{o,max} \frac{C_o}{K_o + C_o} \tag{B.8}$$

Specific growth rate limit:

$$Q_{s,lim} = \frac{\mu_{cr}}{Y_{x/s}^{ox}} \tag{B.9}$$

Oxidative glucose metabolism:

$$Q_{s,ox} = \min \begin{pmatrix} Q_s \\ Q_{s,lim} \\ Y_{s/o}\, Q_{o,lim} \end{pmatrix} \tag{B.10}$$

Reductive glucose metabolism:

$$Q_{s,red} = Q_s - Q_{s,ox} \tag{B.11}$$

Ethanol uptake rate:

$$Q_{e,up} = Q_{e,max} \frac{C_e}{K_e + C_e} \frac{K_l}{K_l + C_s} \tag{B.12}$$

Oxidative ethanol metabolism:

$$Q_{e,ox} = \min \begin{pmatrix} Q_{e,up} \\ (Q_{o,lim} - Q_{s,ox}Y_{o/s})Y_{e/o} \end{pmatrix} \tag{B.13}$$

Ethanol production rate:

$$Q_{e,pr} = Y_{e/s}Q_{s,red} \tag{B.14}$$

Total specific growth rate:

$$\begin{aligned} \mu &= \mu_{ox} + \mu red + \mu e \qquad \text{or} \\ \mu &= Y_{x/s}^{ox}Q_{s,ox} + Y_{x/s}^{red}Q_{s,red} + Y_{x/e}Q_{e,ox} \end{aligned} \tag{B.15}$$

Carbon dioxide production rate:

$$Q_c = Y_{c/s}^{ox}Q_{s,ox} + Y_{c/s}^{red}Q_{s,red} + Y_{c/e}Q_{e,ox} \tag{B.16}$$

Oxygen consumption rate:

$$Q_o = Y_{o/s}Q_{s,ox} + Y_{o/e}Q_{e,ox} \tag{B.17}$$

Respiratory Quotient:

$$RQ = \frac{Q_c}{Q_o} \tag{B.18}$$

The model parameters and initial conditions that are used for dynamic simulations are listed in Table B.1 and Table B.2.

Table B.1. The parameter values of the industrial model.

Parameters	Values
m	0.00321
$K_L a_o$	600
K_e	0.0008
$Y_{c/e}$	0.68
K_l	0.0001
$Y_{o/e}$	1028
K_s	0.002
$Y_{c/s}^{ox}$	2.35
$Q_{e,max}$	0.70805
$Y_{c/s}^{red}$	1.89
$Q_{s,max}$	0.06
$Y_{e/s}$	1.9
$Q_{o,max}$	0.2
$Y_{o/s}$	2.17
μ_{cr}	0.15753
C_o^*	2.41×10^{-4}
$Y_{x/e}$	2.0
C_c^*	0.00001
$Y_{x/s}^{ox}$	4.57063
$K_L a_c$	470.4
$Y_{x/s}^{red}$	0.1
K_o	3×10^{-6}

Table B.2. Initial conditions for dynamic simulation.

State variables	Values
C_s	5×10^{-4}
V	50000
C_e	0
C_o	2.4×10^{-4}
C_c	0
X	0.54

References

1. P. Stanbury, A. Whitaker, and S. Hall, *Principles of fermentation technology.* Oxford: Butterworth-Heinemann, 1995.
2. J. Lee, S. Lee, S.P., and A. Middeelberg, "Control of fed-batch fermentations," *Biotechnology advances,* vol. 17, pp. 29–48, 1999.
3. A. Cinar, S. Parulekar, C. Ündey, and G. Birol, *Batch fermentation: modelling, monitoring, and control.* New York: Marcel Dekker, Inc., 2003.
4. S. Aiba, A. Humphrey, and N. Millis, *Biochemical engineering.* New York: Academic, second ed., 1973.
5. C. o. B. E. National Research Council (U.S.), *Putting biotechnology to work.* National Academy of Science. National Academy Press, 1992.
6. S. Upreti, "A new robust technique for optimal control of chemical engineering processes," *Computers & Chemical Engineering,* 2003. Article in press.
7. K. Y. Rani and V. R. Rao, "Control of fermenters - a review," *Bioprocess Engineering,* vol. 21, pp. 77–88, 1999.
8. K. Ringbom, A. Rothberg, and B. Saxén, "Model-based automation of baker's yeast production," *Journal of Biotechnology,* vol. 51, pp. 73–82, 1996.
9. P. Harms, Y. Kostov, and G. Rao, "Bioprocess monitoring," *Current Opinion in Biotechnology,* vol. 13, pp. 124–127, 2002.
10. A. de Assis and R. Filho, "Soft sensors development for on-line bioreactor state estimation," *Computers & Chemical Engineering,* vol. 24, pp. 1355–1360, 2000.
11. D. Beluhan, D. Gosak, N. Pavlović, and M. Vampola, "Biomass estimation and optimal control of the baker's yeast fermentation," *Computers & Chemical Engineering,* vol. 19, pp. s387–s392, 1995.
12. A. Chéruy, "Software sensors in bioprocess engineering," *Journal of Biotechnology,* vol. 52, pp. 193–199, 1997.
13. R. Gudi, S. Shah, and M. Gray, "Adaptive multirate state and parameter estimation strategies with application to a bioreactor," *American Institute of Chemical Engineering Journal,* vol. 41, pp. 2451–2464, 1995.
14. K. Schügerl, "Progress in monitoring, modeling and control of bioprocesses during the last 20 years," *Journal of Biotechnology,* vol. 85, pp. 149–173, 2001.
15. F. Lei, M. Rotbøll, and S. Jørgensen, "A biochemically structured model for *saccharomyces cerevisiae,*" *Journal of Biotechnology,* vol. 88, pp. 205–221, 2001.

16. M. Di Serio, R. Tesser, and E. Santacesaria, "A kinetic and mass transfer model to simulate the growth of baker's yeast in industrial bioreactors," *Chemical Engineering Journal*, vol. 82, pp. 347–354, 2001.

17. C. Pertev, M. Türker, and R. Berber, "Dynamic modelling, sensitivity analysis and parameter estimation of industrial yeast fermentation," *Computers & Chemical Engineering*, vol. 21, pp. s739–s744, 1997.

18. B. Sonnleitner and O. Käppeli, "Growth of saccharomyces cerevisiae is controlled by its limited respiratory capacity: formulation and verification of a hypothesis," *Biotechnology and Bioengineering*, vol. 28, pp. 927–937, June 1986.

19. R. Berber, C. Pertev, and M. Türker, "Optimization of feeding profile for baker's yeast production by dynamic programming," *Bioprocess Engineering*, vol. 20, pp. 263–269, 1999.

20. Hisbullah, M. Hussain, and K. Ramachandran, "Comparative evaluation of various control schemes for fed-batch fermentation," *Bioprocess and Biosystems Engineering*, vol. 24, pp. 309–318, 2002.

21. M. Ronen, Y. Shabtai, and H. Guterman, "Optimization of feeding profile for a fed-batch bioreactor by an evolutionary algorithm," *Journal of biotechnology*, vol. 97, pp. 253–263, 2002.

22. K. San and G. Stephanopoulos, "Optimization of fed-batch penicillin fermentation: a case of singular optimal control with state constraints," *Biotechnology and Bioengineering*, vol. 28, pp. 72–78, 1989.

23. Y. Pyun, J. Modak, Y. Chang, and H. Lim, "Optimization of biphasic growth of saccharomyces carlsbergensis in fed-batch culture," *Biotechnology and Bioengineering*, vol. 33, pp. 1–10, 1989.

24. M. de Tremblay, M. Perrier, C. Chavarie, and J. Archambault, "Optimization of fed-batch culture of hybridoma cells using dynamic programming: single and multi feed cases," *Bioprocess Engineering*, vol. 7, pp. 229–234, 1992.

25. J. Lee, H. Lim, Y. Yoo, and Y. Park, "Optimization of feed rate profile for the monoclonal antibody production," *Bioprocess Engineering*, vol. 20, pp. 137–146, 1999.

26. R. King, *Computational intelligence in control engineering*. Control Engineering Series, New York: Marcel Dekker, Inc., 1999.

27. A. Zilouchian and M. Jamshidi, *Intelligent control systems using soft computing methodologies*. Florida, USA: CRC Press LLC, 2001.

28. R. Kamimura, K. Konstantinov, and G. Stephanopoulos, "Knowledge-based systems, artificial neural networks and pattern recognitions: applications to biotechnological processes," *Current Opinion in Biotechnology*, vol. 7, pp. 231–234, 1996.

29. P. Patnaik, "Artificial intelligence as a tool for automatic state estimation and control of bioreactors," *Laboratory Robotics and Automation*, vol. 9, pp. 297–304, 1997.

30. M. Karim, T. Yoshida, S. Rivera, V. Saucedo, B. Eikens, and G. Oh, "Global and local neural network models in biotechnoloy: application to different cultivation processes," *Journal of Fermentation and Bioengineering*, vol. 83, no. 1, pp. 1–11, 1997.

31. M. Hussain, "Review of the applications of neural networks in chemical process control - simulation and online implementation," *Artificial Intelligence in Engineering*, vol. 13, pp. 55–68, 1999.

32. P. Fleming and R. Purshouse, "Evolutionary algorithms in control systems engineering: a survey," *Control Engineering Practice*, vol. 10, pp. 1223–1241, 2002.

33. J. Almeida, "Predictive non-linear modelling of complex data by artificial neural network," *Current Opinion in Biotechnology*, vol. 13, pp. 72–76, 2002.

34. M. Hagan, H. Demuth, and M. Beale, *Neural network design*. PWB publishing company, 1996.

35. R. Żbikowski and A. Dzieliński, "Neural approximation: a control perspective," in *Neural network engineering in dynamic control systems* (K. Hunt, G. Irwin, and K. Warwick, eds.), Advances in industrial control Series, pp. 1–25, London: Springer-Verlag, 1995.

36. G. Ng, *Application of neural networks to adaptive control of nonlinear systems*. UMIST Control System Center Series, England: Research Studies Press Ltd., 1997.

37. D. Mandic and J. Chambers, *Recurrent neural networks for prediction*. New York: John Wiley & Soms, Ltd., 2001.

38. J. Elman, "Finding structure in time," *Cognitive Science*, vol. 14, pp. 179–211, 1990.

39. P. Patnaik, "A recurrent neural network for a fed-batch fermentation with recombinant escherichia coli subject to inflow disturbances," *Process Biochemistry*, vol. 32, no. 5, pp. 391–400, 1997.

40. L. Valdez-Castro, I. Baruch, and J. Barrera-Cortés, "Neural networks applied to the prediction of fed-batch fermentation kinetics of bacillus thuringiensis," *Bioprocess and Biosystems Engineering*, vol. 25, pp. 229–233, 2003.

41. Y. I, W. Wu, and Y. Liu, "Neural network modelling for on-line state estimation in fed-batch culture of l-lysine production," *The Chemical Engineering Journal*, vol. 61, pp. 35–40, 1996.

42. L. Chen, S. Nguang, X. Li, and X. Chen, "Soft sensors for on-line biomass measurements," *Bioprocess and Biosystems Engineering*, vol. 26, no. 3, pp. 191–195, 2004.

43. C. Komives and R. Parker, "Bioreactor state estimation and control," *Current Opinion in Biotechnology*, vol. 14, pp. 468–474, 2003.

44. B. Schenker and M. Agarwal, "Predictive control of a bench-scale chemical reactor based on neural-network models," *IEEE Transaction on control systems technology*, vol. 6, pp. 388–400, 1998.

45. K. Zuo and W. Wu, "Semi-realtime optimization and control of a fed-batch fermentation system," *Computers & Chemical Engineering*, vol. 24, pp. 1105–1109, 2000.

46. L. Zorzetto, R. M. Filho, and M. Woif-Maciel, "Process modelling development through artificial neural networks and hybrid models," *Computers & Chemical Engineering*, vol. 24, pp. 1355–1360, 2000.

47. C. Alippi and V. Piuri, "Experimental neural networks for prediction and identification," *IEEE Transactions on instrumentation and measurement*, vol. 45, pp. 670–676, 1996.

48. B. Chaudhuri and J. Modak, "Optimization of fed-batch bioreactor using neural network model," *Bioprocess Engineering*, vol. 19, pp. 71–79, 1998.

49. P. Lednický and A. Mészáros, "Neural network modelling in optimization of continuous fermentation processes," *Bioprocess Engineering*, vol. 18, pp. 427–432, 1998.

50. T. Chow and X. Li, "Modeling of continuous time dynamical systems with input by recurrent networks," *IEEE Transactions on circuits and systems - I: Fundamental theory and applications*, vol. 47, pp. 575–578, 2000.

51. J. Zhang, A. Morris, and E. Martin, "Long-term prediction models based on mixed order locally recurrent neural networks," *Computers & Chemical Engineering*, vol. 22, no. 7, pp. 1051–1063, 1998.

52. D. Pham and S. Oh, "Identification of plant inverse dynamics using neural networks," *Artificial Intelligence in Engineering*, vol. 13, pp. 309–320, 1999.

53. R. Hernández, J. Gallegos, and J. Reyes, "Simple recurrent neural network: A neural network structure for control systems," *Neurocomputing*, vol. 23, pp. 277–289, 1998.

54. P. Teissier, B. Perret, E. Latrille, J. Barillere, and G. Corrieu, "A hybrid recurrent neural network model for yeast production monitoring and control in a wine base medium," *Journal of Biotechnology*, vol. 55, pp. 157–169, 1997.

55. A. Shaw, F. D. III, and J. Schwaber, "A dynamic neural network approach to nonlinear process modeling," *Computers & Chemical Engineering*, vol. 21, no. 4, pp. 371–385, 1997.

56. T. Becker, T. Enders, and A. Delgado, "Dynamic neural networks as a tool for the online optimization of industrial fermentation," *Bioprocess and Biosystems Engineering*, vol. 24, pp. 347–354, 2002.

57. G. Acuña, E. Latrille, C. Béal, and G. Corrieu, "Static and dynamic neural network models for estimating biomass concentration during thermophilic lactic acid bacteria batch cultures," *Journal of Fermentation and Bioengineering*, vol. 85, no. 6, pp. 615–622, 1998.

58. Y. Tian, J. Zhang, and J. Morries, "Optimal control of a fed-batch bioreactor based upon an augmented recurrent neural network model," *Neurocomputing*, vol. 48, pp. 919–936, 2002.

59. J. Holland, *Adaptation in natural and artificial systems*. Ann Arbor, MI: The University of Michigan Press, 1975.

60. T. Back, *Evolutionary algorithms in theory and practice: evolution strategies, evolutionary programming, genetic algorithms*. New York: Oxford University Press, 1996.

61. D. Fogel, *Evolutionary computation: towards a new philosophy of machine intelligence*. Piscataway, NJ: IEEE Press, 1995.

62. D. Goldberg, *Genetic algorithms in search, optimization and machine learning*. Addision-Wesley, 1989.

63. J. Koza, *Genetic programming*. Cambridge, MA: The MIT Press, 1992.

64. Z. Michalewicz, *Genetic algorithm + data structures = evolution programs*. Berlin: Springer-Verlag, 1996.

65. M. Mitchell, *An introduction to genetic algorithms*. Cambridge, MA: The MIT Press, 1996.

66. M. Tomassini, "A survey of genetic algorithms," in *Annual reviews of computational physics* (D. Stauffer, ed.), vol. III, pp. 87–118, World Scientific, 1995.

67. L. Davis, *Handbook of genetic algorithms*. New York: Van Nostrand Reinhold, 1991.

68. C. Reeves and J. Rowe, *Genetic algorithms: principles and perspectives*. USA: Kluwer Academic Publishers, 2003.

69. M. Jamshidi, L. dos Santos Coelho, R. Krohling, and P. Fleming, *Robust control systems with genetic algorithms*. Boca Raton: CRC Press, 2003.

70. L. Zadeh, "The evolution of systems analysis and control: a personal perspective," *IEEE Control Systems Magazine*, vol. 16, no. 3, pp. 95–98, 1996.

71. C. Di Massimo, G. Montague, M. Willis, M. Tham, and A. Morris, "Towards improved penicillin fermentation via artificial neural networks," *Computers & Chemical Engineering*, vol. 16, no. 4, pp. 283–291, 1992.

72. S. Linko, Y. Zhu, and P. Linko, "Applying neural networks as software sensors for enzyme engineering," *Trends in Biotechnology*, vol. 17, pp. 155–162, 1999.

73. P. Dacosta, C. Kordich, D. Williams, and J. B. Gomm, "Estimation of inaccessible fermentation states with variable inoculum sizes," *Artificial Intelligence in Engineering*, vol. 11, pp. 383–392, 1997.

74. L. Fortuna, A. Rizzo, M. Sinatra, and M. Xibilia, "Soft analyzers for a sulfur recovery unit," *Control Engineering Practice*, 2003. Article in press.

75. J.-G. Na, Y. Chang, B. Chung, and H. Lim, "Adaptive optimization of fed-batch culture of yeast by using genetic algorithms," *Bioprocess and Biosystems Engineering*, vol. 24, pp. 299–308, 2002.

76. S. Nguang, L. Chen, and X. Chen, "Optimisation of fed-batch culture of hybridom cells using genetic algorithms," *ISA Transaction*, vol. 40, pp. 381–389, 2001.

77. D. Sarkar and J. Modak, "Optimization of fed-batch bioreactors using genetic algorithms," *Chemical Engineering Science*, vol. 58, pp. 2283–2296, 2003.

78. P. Patnaik, "Hybrid neual simulation of a fed-batch bioreactor for a nonideal recombinant fermentation," *Bioprocess and Biosystems Engineering*, vol. 24, pp. 151–161, 2001.

79. A. Zalzala and P. Fleming, "Genetic algorithms in engineering systems," in *IEE Control Engineering Series*, vol. 55, 1997. ISBN 0852969023.

80. J. Roubos, G. van Straten, and A. Boxtel, "An evolutionary strategy for fed-batch bioreactor optimization; concepts and performance," *Journal of biotechnology*, vol. 67, pp. 173–187, 1999.

81. M. Ranganath, S. Renganathan, and C. Gokulnath, "Identification of bioprocesses using genetic algorithm," *Bioprocess Engineering*, vol. 21, pp. 123–127, 1999.

82. L. Chen, S. Nguang, and X. Chen, "On-line identification and optimization of feed rate profiles for high productivity fed-batch culture of hybridoma cells using genetic algorithms," *ISA Transactions*, vol. 41, pp. 409–419, October 2002.

83. M. Lim, Y. Tayeb, J. Modak, and P. Bonte, "Computational algorithms for optimal feed rates for a class of fed-batch fermentation: Numerical results for penicillin and cell mass production," *Biotech. Bioeng.*, vol. 28, pp. 1408–1420, 1986.

84. V. Summanwar, V. Jayaraman, B. Kulkarni, H. Kusumakar, K. Gupta, and J. Rajesh, "Solution of constrained optimization problems by multi-objective genetic algorithm," *Computers & Chemical Engineering*, vol. 26, pp. 1481–1492, 2002.

85. Y. Liu, F. Wang, and W. Lee, "On-line monitoring and controlling system for fermentation process," *Biochemical Engineering Journal*, vol. 7, pp. 17–25, 2001.

86. L. Chen, S. Nguang, X. Li, and X. Chen, "Soft sensor development for on-line biomass estimation," in *Proc. Int. Conf. on control and automation*, (Xiamen, China), pp. 1441–1445, June 2002. IEEE Catalog No. 02EX560C, ISBN 0-7803-7413-4.

87. J. Gomes and A. Menawat, "Precise control of dissolved oxygen in bioreactors - a model-based geometric algorithm," *Chemical Engineering Science*, vol. 55, pp. 67–78, 2000.

88. C. Lee, S. Lee, K. Jung, S. Katoh, and E. Lee, "High dissolved oxygen tension enhances heterologous protein expression by recombinant pichia pastoris," *Process Biochemistry*, vol. 38, pp. 1147–1154, 2003.

89. Z. Nor, M. Tamer, J. Scharer, M. Moo-Young, and E. Jervis, "Automated fed-batch culture of kluyveromyces fragilis based on a novel method for on-line estimation of cell specific growth rate," *Biochemical Engineering Journal*, vol. 9, pp. 221–231, 2001.

90. M. Åkesson, E. Karlsson, P. Hagander, J. Axelsson, and A. Tocaj, "On-line detection of acetate fromation in escherichia coli cultures using dissolved oxygen responses to feed transients," *Biotechnology and Bioengineering*, vol. 64, pp. 590–598, September 1999.

91. R. Williams and D. Zipser, "A learning algorithm for continually running fully recurrent neural networks," *Neural Computation*, vol. 1, pp. 270–280, 1989.

92. R. Berber, C. Pertev, and M. Tarker, "Optimization of feeding profile for baker's yeast production by dynamic programming," *Bioprocess Engineering*, vol. 20, pp. 263–269, 1998.

93. L. Scales, *Introduction to non-linear optimization*. New York: Springer-Verlag, 1985.

94. R. Simutis and A. Lubbert, "Exlotoratory analysis of bioprocesses using artificial neural-network-based methods," *Biotechnol Prog*, vol. 13, pp. 479–487, 1997.

95. G. Montague, A. Morris, and M. Tham, "Enchancing bioprocess operability with generic software sensors," *J Biotechnol*, vol. 25, pp. 183–204, 1992.

96. Q. Chen and W. Weigand, "Dynamic optimization of nonlinear process by combining a neural net model with udmc," *AIChE J.*, vol. 40, pp. 1488–1497, 1994.

97. B. Atkinson and F. Mavituna, *Biochemical Engineering and Biotechnology Handbook*. New York: Stockton Press, second ed., 1991.

98. C. de Boor, *A practical guide to splines*. New York: Springer-Verlag, revised ed., 2001.

99. J. Peres, R. Oliveira, and S. F. de Azevedo, "Knowledge based modular networks for process modelling and control," *Computers & Chemical Engineering*, vol. 25, pp. 783–791, 2001.

100. S. Park and W. Ramirez, "Optimal production of secreted protein in fed-batch reactors," *A.I.Ch.E.J.*, pp. 1550–1558, 1988.

101. L. Nurmianto, P. Greenfield, and P. Lee, "Optimal feeding policy for recombinant protein production via the yeast *Saccharomyces cerevisiae* in fed-batch culture," *Process Biochemistry*, vol. 29, pp. 55–68, 1994.

102. J. van Impe and G. Bastin, "Optimal adaptive control of fed-batch fermentation processes," *Control Engineering Practice*, vol. 3, no. 7, pp. 939–954, 1995.

103. Y. Huang, G. Reklaitis, and V. Venkatasubramanian, "Model decomposition based method for solving general dynamic optimization problems," *Computers & Chemical Engineering*, vol. 26, pp. 863–873, 2002.

104. D. Sridhar, E. Bartlett, and R. Seagrave, "Information theoretic subset selection for neural network models," *Computers & Chemical Engineering*, vol. 22, pp. 613–626, 1998.

105. B. Lennox, P. Rutherford, G. Montague, and C. Haughin, "Case study investigation the application of neural network for process modelling and condition monotoring," *Computers & Chemical Engineering*, vol. 22, no. 11, pp. 1573–1579, 1998.

106. H. te Braake, H. van Can, and H. Verbruggen, "Semi-mechanistic modelling of chemical processes with neural networks," *Engineering Application of Artificial Intelligence*, vol. 11, pp. 507–515, 1998.

107. M. Iatrou, T. Berger, and V. Marmarelis, "Modelling of nonlinear nonstationary dynamic systems with a novel class of artificial neural networks," *IEEE transactions on neural networks*, vol. 10, March 1999.

108. R. Simutis, R. Oliveira, M. Manikowski, S. F. de Azevedo, and A. Lübbert, "How to increase the performance of models for process optimization and control," *Journal of Biotechnology*, vol. 59, pp. 73–89, 1997.

109. G. Walker, *Yeast physiology and biotechnology*. New York: John Wiley & Sons, Inc., 1998.

110. V. Saucedo and M. Karim, "Analysis and comparison of input-output models in a recombinant fed-batch fermentation," *Journal of Fermentation and Bioengineering*, vol. 83, no. 1, pp. 70–78, 1997.

111. D. Seborg, T. Edgar, and D. Mellichamp, *Process dynamics and control*. New York: John Wiley & Sons, 1989.

112. L. Chen, S. Nguang, X. Li, and X. Chen, "Optimal control of a fed-batch bioreactor using genetic algorithm based on cascade dynamic neural network model," in *Proc. Int. Conf. on automatic control*, (Auckland, New Zealand), December 2002.

113. P. Koprinkova and M. Petrova, "Data-scaling problems in neural-network training," *Engineering Application of Artificial Intellingence*, vol. 12, pp. 281–296, 1999.

114. S. Valentinotti, B. Srinivasan, U. Holmberg, D. Bonvin, C. Cannizzaro, M. Rhiel, and U. von Stockar, "Optimal operation of fed-batch fermentation via adaptive control of overflow metabolite," *Control Engineering Practice*, vol. 11, pp. 665–674, 2003.

115. M. Åkesson, P. Hagander, and J. Axelsson, "Probing control of fed-batch cultivations: analysis and tuning," *Control Engineering Practice*, vol. 9, pp. 709–723, 2001.

116. I. Smets, K. Versyck, and J. V. Impe, "Optimal control theory: a generic tool for identification and control of (bio-)chemical reactors," *Annu. Rev. Contr.*, vol. 26, pp. 57–73, 2002.

117. D. Levisauskas, V. Galvanauskas, S. Henrich, K. Wilhelm, N. Volk, and A. Lübbert, "Model-based optimization of viral capsid protein production in fed-batch culture of recombinant *escherichia coli*," *Bioprocess and Biosystems Engineering*, vol. 25, pp. 255–262, 2003.

118. J. Nielsen and J. Villadsen, *Bioreactor engineering principles*. New York: Plenum Press, 1994.

119. H. te Braake, R. Babuska, E. van Can, and C. Hellinga, "Predictive control in biotechenology using fuzzy and neural models," in *Advanced Instrumentation, Data Interpretation, and Control of Biotechnological Processes* (I. D. Dordrecht, ed.), Kluwer Academic Publisher, 2000.